SPATIAL
CLUSTER
MODELLING

SPATIAL CLUSTER MODELLING

Edited by

Andrew B. Lawson
Department of Mathematical Sciences
University of Aberdeen
Aberdeen, UK

David G.T. Denison
Department of Mathematics
Imperial College of Science, Technology and Medicine
London, UK

CRC Press
Taylor & Francis Group
Boca Raton London New York

CRC Press is an imprint of the
Taylor & Francis Group, an **informa** business

A CHAPMAN & HALL BOOK

First published 2002 by Chapman and Hall

Published 2019 by CRC Press
Taylor & Francis Group
6000 Broken Sound Parkway NW, Suite 300
Boca Raton, FL 33487-2742

First issued in paperback 2019

No claim to original U.S. Government works

ISBN-13: 978-0-367-45491-3 (pbk)
ISBN-13: 978-1-58488-266-4 (hbk)

Visit the Taylor & Francis Web site at
http://www.taylorandfrancis.com

and the CRC Press Web site at
http://www.crcpress.com

Library of Congress Card Number 2002019297

Library of Congress Cataloging-in-Publication Data

Lawson, Andrew (Andrew B.)
    Spatial cluster modeling / edited by Andrew B. Lawson, David G.T. Denison.
        p. cm.
    Includes bibliographical references and index.
    ISBN 1-58488-266-2 (alk. paper)
    1. Cluster analysis. 2. Spatial analysis. I. Denison, David G. T. II. Title.

QA278 .L39 2002
519.5′3—dc21   2002019297

Contents

List of Contributors

Adrian Baddeley	University of Western Australia
Dankmar Böhning	Free University of Berlin
Simon D. Byers	AT&T Labs, New Jersey
Allan B. Clark	University of Aberdeen
Murray K. Clayton	University of Wisconsin, Madison
David G.T. Denison	Imperial College, London
José Tomé A.S. Ferreira	Imperial College, London
Jürgen Gallinat	Free University of Berlin
Ronald B. Gangnon	University of Wisconsin, Madison
Fred Godtliebsen	University of Tromsø
Christopher M. Hans	Duke University
Christopher C. Holmes	Imperial College, London
John T. Kent	University of Leeds
Hyoung Moon Kim	Texas A&M University
Andrew B. Lawson	University of Aberdeen
Marc Loizeaux	Florida State University, Tallahassee
Ian W. McKeague	Florida State University, Tallahassee

Bani K. Mallick   Texas A&M University

Kanti V. Mardia   University of Leeds

J. Steve Marron   University of North Carolina, Chapel Hill

Jesper Møller   Aalborg University

Stephen M. Pizer   University of North Carolina, Chapel Hill

Adrian E. Raftery   University of Washington, Seattle

Peter Schlattmann   Free University of Berlin

David A. van Dyk   Harvard University

Marie-Colette van Lieshout   CWI, Amsterdam

Rasmus P. Waagepetersen   Aalborg University

Christopher K. Wikle   University of Missouri–Columbia

Preface

The development of statistical methodology for the analysis of spatial data has seen considerable advances since the publication of the seminal work of Cressie (1993). In particular, the development of fast computational algorithms for sampling of complex Bayesian models (most notably Markov chain Monte Carlo algorithms) has allowed a wide range of problems to be addressed which hitherto could not be directly analysed. Many spatial problems can be considered within the paradigm of hierarchical Bayesian modelling and so the emphasis within this volume will lie within that area.

The aim of this volume is not to present a general review of spatial statistical modelling but rather to focus on the area of spatial *cluster* modelling. Hence the theme of this work is the highlighting of the diverse approaches to the definition of clusters and clustering in space (and its adjunct space-time), and to present state-of-the-art coverage of the diverse modelling approaches which are currently available. In Chapter 1 we provide a brief historical introduction to the subject area and, in particular, compare conventional and spatial clustering. In addition this chapter introduces the notation and different areas of study explored. After this initial chapter the volume is split into 3 parts, each relating to a specific area of cluster modelling. Part I deals with point and object process modelling, Part II involves spatial process modelling, while Part III contains papers relating to spatio-temporal models.

One of the features of modelling spatial data is the need to use fast computational algorithms to be able to evaluate the complex posterior distributions or likelihood surfaces which arise in spatial applications. The 1980s saw the development of Markov chain Monte Carlo algorithms based on the Gibbs and Metropolis-Hastings samplers, and witnessed rapid development of models for complex spatial problems. Not only could existing models be sampled from but newer more sophisticated models could also be developed and applied. Often these models are of a hierarchical form so this naturally leads to the Bayesian paradigm being of importance in a great deal of the work.

As the potential fields of application for spatial methods are so wide we cannot hope to cover all of them. Nevertheless the chapters here do make reference to data in astrophysics (Chapter 10), spatial epidemiology (Chapters 5,7,8,14), ecology (Chapters 4,11), imaging (Chapter 13), geology and the geosciences (Chapters 6,4,7,9,12). In addition, the volume provides a useful insight into the current issues and methodology used for spatial cluster modelling. We have specifically included the burgeoning area of spatio-temporal modelling as an important extension to standard spatial data analysis and Chapters 12,13,14 specifically deal with this topic.

Finally we would like to thank all the contributors for their timely and thoughtful articles. In addition, we acknowledge the help of the staff at CRC Press, in particular Kirsty Stroud, Jasmin Naim and Helena Redshaw for their continued support and encouragement in the production of this volume.

CHAPTER 1

Spatial Cluster Modelling: An Overview

A.B. Lawson   *D.G.T. Denison*

1.1 Introduction

When analysing spatial data one is often interested in detecting deviations from the expected. For instance, we may be interested in the answers to questions like, "Is there an unusual aggregation of leukemia cases around a nuclear power station ?" or, "Where is it likely that the air pollution level is above the legally allowed limit ?". In both cases the focus is on finding regions in (usually two-dimensional) space in which higher than expected counts, or readings, are observed. We shall call such areas *clusters* and determining their nature forms the focus of this work.

This volume brings together a collection of papers on the topic of spatial cluster modelling and gives descriptions of various approaches which begin to solve the problem of detecting clusters. The papers are statistical in nature but draw on results in other fields as diverse as astrophysics, medical imaging, ecology and environmental engineering.

Two examples of the sort of spatial processes that we shall consider here are displayed in Figs. 1.1-1.2. Fig. 1.1 is an example of a point process, where each dot is an "event" (in this case the occurrence of a cancer). Here it is of interest to determine whether the cases are more aggregated, or *clustered*, than expected and whether the clustering relates to the locations of any possible pollution sources. To assess this a background control disease map (which is not shown) is often used to represent the expected variation in the distribution of cases; this is often a function of the relative population density.

Fig. 1.2 is an example of a dataset that consists of observations of an underlying spatial process at a number of locations. The usual aim of an analysis of this type of data is to determine the value of the spatial process at all the locations in the domain of interest, assuming that each measurement is only observed in the presence of a random error component. However, we may also be interested in determining areas where the process is above some predefined limit or even, in some sense, above average.

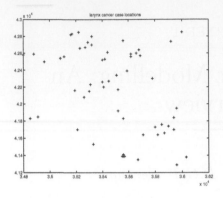

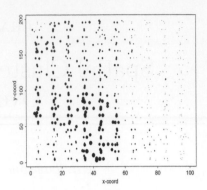

Figure 1.1 *An example of a point process: The Larynx cancer case event map (Diggle 1990) relating to cases in Lancashire, UK from 1974 to 1983.*

Figure 1.2 *An example of observations of a spatial process: The Piazza Road data (Higdon et al. 1999). Each dot represents a location where a measurement of the dioxin concentration in an area around the Piazza Road is taken. The size of the dots gives an indication of the observed measurements, with larger dots representing higher concentrations.*

Spatio-temporal data will also be looked at in this volume. In these examples either measurements or point processes are observed over time and similar questions arise but now clusters can occur in space, in time, or even in space and time jointly.

The analysis of spatial clustering has had a varied history with developments often resulting from particular applications, and so these developments have related to varying interest over time in different applications. For example, much work in the 1960s and 1970s on clustering developed from ecology applications (e.g. Diggle 1983, Ripley 1981), whereas an increased interest in image analysis in the 1980s led to associated advances in image segmentation and object recognition (e.g. Besag et al. 1991). Increased public and scientific interest in environmental hazards and health in the late 1980s and 1990s, has led to increased emphasis on cluster detection in small area health studies (Lawson et al. 1999, Elliott et al. 1999, Lawson 2001). The context of this historical development in relation to methodological progress is discussed more fully in the next section.

1.2 Historical Development

The analysis of clustering has a long history in statistical science. In one dimension the analysis of aggregation of data around a preferred location is at the heart of much statistical work, whether the focus is on the mean tendency of the aggregation of data or on its spread or variance. In addition in cluster studies the location of the maximum aggregation may also be of interest (modal property). In two dimensions, the natural extension of these ideas is to a two-dimensional aggregation, perhaps around a single point. In this case the centre of the aggregation may be defined either by mean or modal properties while the variance or spread of the aggregation can be defined around the putative centre which is now a two-dimensional location. In the case of spatial data these quantities have obvious interpretations as cluster centre location and cluster spread.

In the spatial domain, it is possible to view a clustered pattern in different ways depending on the focus of the analysis. First, it may be possible to conceive of the pattern as the realisation of a random process which produces aggregations as a result of the global structure of the process, whereby a small number of parameters control the scale and frequency of aggregations but the defined process does not parameterise the *locations* of the aggregations. This is akin to the geostatistical view of random processes, where the intensity or local density of events is defined by, for example, a spatial Gaussian process. The peaks of this process would correspond with local aggregations, but no parameterisation of the locations is made. Recent examples of this approach can be found in Cressie (1993) and Diggle et al. (1998). Essentially, this approach regards the aggregations as produced by random effects which are governed by global model parameters, i.e. the degree of aggregation and spread of the aggregations would be controlled by a small number of global parameters. This form of random effect modelling is at the heart of much hierarchical Bayesian modelling in this context, and in the literature of spatial applications the term *clustering* or *cluster modelling* is used to refer to such random effect modelling. An example from small area health studies is Clayton and Bernardinelli (1992).

The second approach to the modelling of clusters is to include within the modelled process specific elements which relate to cluster location and how these locations relate to the surrounding data. Much of the early work in stochastic geometry relating to point processes examined clustering of point processes and models such as the Neyman-Scott and Cox cluster processes. The fundamental feature of these processes was the definition of a set of cluster centres around which the *offspring* (data) lie. The term offspring comes from the idea that the clustering could arise from a multigenerational process. The variation in the local aggregation of data is thought to be summarised by the local density of events around the cluster centre set. In the case of a Neyman-Scott or Poisson cluster process the local density is a function of distance from the cluster centre set within an infinite plane. The appearance of such clustered event sets is controlled by the location of centres and the spread of the distribution around these centres.

The dichotomy between these two approaches leads from the different focus and objectives of the analyses. While both approaches use global parameters to describe the local variation in space, they are distinguished by their *specificity* with regard to cluster location. Within this volume a range of approaches to clustering is described including examples from both types of analysis and some where the distinctions are more blurred.

Most of the earliest work in clustering related to the development of theoretical models for clusters and cluster processes, with limited application of the models to real data. In addition, the earliest theoretical work concerned one-dimensional processes. Early reference sources for such models and their properties are Snyder (1975) and Cox and Isham (1980). In the 1970s, increased attention was paid to the application of cluster models to ecological and biological examples. In these developments, summary statistics were developed which would allow the assessment of first second order properties of realisations of point processes. The Ripley-K function, nearest neighbour and inter-event statistics are all examples of these developments (see Diggle 1983). Such methods have been further developed in applications to processes of objects other than points (e.g. Baddeley 1999).

During the 1980s, developments in the analysis of pixellated images led to the application of Markov random field (MRF) models. These models were intended for segmentation or noise reduction in images. These models were often characterised by the inclusion of a neighbourhood effect, often included within a prior distribution for the image. Bayesian hierarchical modelling of images was a natural development as prior beliefs about the spatial structure of the image could be incorporated within a spatial prior distribution, while measurement error could be included within the likelihood.

Associated with these developments, were the increased use of stochastic algorithms for maximising or sampling posterior distributions of image features – in particular, the development of Gibbs and Metropolis-Hastings samplers (e.g. Gilks et al. 1996) and simulated annealing for maximisation (Ripley 1988). These developments regard aggregation as described by non-specific neighbourhood structures with associated global parameters. As such they have similarities to geostatistical models where continuous variation is modelled by predefined covariance structures. Essentially these are random effect models.

At the other end of the spectrum during the 1980s and 1990s, another form of image modelling developed which emphasised the specific locational features of the image. This approach can be described as object recognition or object modelling. In this collection of approaches the locational characteristics of the image features are specifically modelled (e.g. the landmark methods of Dryden and Mardia 1997). Object process modelling is one area which has seen considerable development. Here a noisy image is assumed to have an underlying distribution of objects. The aim is to reconstruct the noise-free object set. Often for simplicity the objects are assumed to be simple geometric shapes and have associated with them a specific point which fixes their location. Simple processes could consist of lines, circles or even tessellations and triangulations.

The recovery of such processes from images has seen the development and application of sophisticated Markov Chain Monte Carlo (MCMC) (Baddeley and van Lieshout 1993, Cressie and Lawson 2000) culminating in the use of special birth-death MCMC algorithms (Geyer and Møller 1994, Stephens 2000).

The above two approaches to image analysis have many variants which consist of either components of random effect and object recognition tasks. For example some models attempt to group or classify areas of images, and these *partition* models can be thought of as cluster models with locational components. A review of these different approaches is found in Hurn et al. (2001).

While image analysis has seen the greatest development of novel methodologies, other application areas have also seen considerable progress in recent years, particularly in the application of Bayesian hierarchical models. Such areas as climatology, geosciences, environmetrics, genetics and spatial epidemiology have seen considerable advances in the modelling of clustered data. Within climatology and geo-/environmental sciences the development of geostatistical methods related to kriging and Bayesian hierarchical models for random effects has been marked (e.g. Wikle et al. 1998, Mardia et al. 1998).

In the last decade a considerable body of work has developed with a focus on small area health data, mainly in the area of spatial epidemiology. The study of disease clustering has a considerable history and many approaches have been advanced ranging from hypothesis testing for clusters to more recently non-specific (random effect) and specific cluster modelling (for a review see Chapter 6 of Lawson 2001). Model-based approaches to clustering of diseases has developed both in terms of random effect modelling of clustering (Clayton and Bernardinelli 1992, Breslow and Clayton 1993, Bernardinelli et al. 1995, Waller et al. 1997a, Zia et al. 1997, Knorr-Held and Besag 1998, Langford et al. 1999, Diggle et al. 1998, Ghosh et al. 1998, Best et al. 1998, Byers and Besag 2000, Knorr-Held 2000, Sun et al. 2000, Best and Wakefield 1999, Lawson and Clark 2001) and, to a lesser extent, in specific modelling of clusters and their locations (Lawson and Clark 1999a, Lawson and Clark 1999b, Gangnon and Clayton 2000). Variants of these two extremes of random effect and random object modelling have been proposed. For example, partition models which group small areas into constant risk classes have been suggested as a form of cluster modelling (Schlattmann and Böhning 1993, Knorr-Held and Rasser 2000). On the other hand, in disease mapping it may be beneficial to assume little prior information about parametric forms of cluster, and so nonparametric approaches may also be useful (e.g. Kelsall and Diggle 1995a, Kelsall and Diggle 1998).

In what follows we reflect this spectrum of approaches between random effect and random object modelling approaches by presenting a review of the range of approaches currently advocated for cluster modelling. We also include nonparametric approaches which provide a weakly model-based alternative to the parameterised approaches discussed above.

1.2.1 Conventional Clustering

Cluster analysis in conventional non-spatial statistical applications has developed in parallel with its spatial counterpart but has a longer pedigree. In the following, we seek to draw parallels with conventional cluster analysis and to show how spatial approaches closely relate to a subset of non-spatial clustering methods. A recent review of the field of non-spatial cluster analysis is given in Everitt et al. (2001). The basis of cluster analysis is the notion of classification of objects into categories based on a measure of similarity. Often classification is a task which allows the differentiation of sub-populations at an early stage of a subject's development, and indeed cluster analysis is often applied to yield information about groupings or aggregations in the data at an exploratory stage in an analysis.

Before considering the basic parametric forms of clustering it should be noted that exploratory clustering analysis of multi-dimensional data is often performed by using nonparametric density estimation or related techniques (see the relevant chapters of Everitt et al. 2001, Duda et al. 2001). These methods are also applicable where the domain is spatial or spatio-temporal. An application of these methods is discussed here in Chapter 2, and examples of their use in epidemiological applications are given in Bithell (1990), Kelsall and Diggle (1995b).

Bayesian decision theory can be applied to simple classification problems where measurements on a range of variables are used to classify data into particular groups. For example, in medical diagnoses a range of physical and biomedical measurements is made on patients and these are used to classify the patient (in the simple binary-state model) into "having a disease", or not. Often in this case a simple belief network and naive Bayesian classifier can be applied (Lucas 2001). More complex Bayesian approaches can be used for multi-state models and Bayesian discrimination can approximate the behaviour of neural network classification algorithms. Many of these algorithms use information about the *labelling* (or classification) of data items, using some training data, i.e. in the medical diagnosis problem the distributions of measurements/symptoms for a particular disease are known from historical data and these can be used to train the classifier.

In clustering studies the aim is usually to examine data which do not contain labels and it is the focus of the analysis to find groupings, or clusters, in the data. This is a form of unsupervised learning, where no training data are used (see Chapter 10 of Duda et al. 2001). In the case of clustering, the focus is to assign labels to the data, which represent the clusters in the data. Often the number of clusters will be unknown as well as the labelling. In addition, it may be important to estimate some characteristic feature of the cluster. For example, it may be hypothesised that each cluster has a centre around which the data aggregate. Hence the emphasis in classification is to recover the labels of data, whereas the object of clustering is often to assess the number and location and/or scale of cluster features. An example would be the estimation of the number and location of galaxies in an astrophysical example. The labelling of stars with galaxy number could be a secondary task. However, in many forms of analysis the labelling and estimation of cluster features are related and can be included within iterative algorithms for cluster feature extraction. For example the conditional distribution of galaxy number and location could be specified in relation to a given labelling of stars, and in turn the labelling of stars could be conditionally related to the number and location of galaxies.

Usually clustering algorithms are based on a metric which describes how closely related respective data items are to each other, or to unobserved cluster features. These metrics can be distance-based or take other forms (see Chapter 3 of Everitt et al. 2001). In addition to this, such algorithms use a clustering criterion to establish the appropriate grouping or labelling of the data.

A fundamental form of model which can be used for cluster problems is that of the mixture distribution. The classical cluster mixture assumes that the data come from a number of clusters, m, where the prior probability of each cluster is known along with the forms of cluster-conditional probability densities. However, the values of the cluster feature vector and the total number of clusters m are unknown. We define a data vector by x, the feature vector by $\theta = (\theta_1, \ldots, \theta_m)$, the prior cluster probability as $p(\omega_j)$ $(j = 1, \ldots, m)$ and the jth cluster-conditional probability densities as $p(x|\omega_j, \theta_j)$. Then the probability density function of the data vector is

$$p(x|\theta) = \sum_{j=1}^{m} p(x|\omega_j, \theta_j)p(\omega_j). \tag{1.1}$$

Usually the focus is on estimation of the feature vector θ. For example, these could be means and variances of Gaussian densities, or could be other relevant feature parameters. If classification is the goal, then once θ is known the mixture can be decomposed into its components and a maximum *a posteriori* classifier could be used. If the component densities are Gaussian then it is possible for mixture components to overlap and hence the clustering criterion does not restrict the cluster form to be disjoint. Other mixture forms produce disjoint non-overlapping partitions of the space.

The above brief discussion focuses on what is, in computer science terminology, a 'flat' classification: that is, the clusters exist within one level of organisation only. Flat classifications could be disjoint (such as discriminant functions) or overlapping (such as mixture densities). However it is also possible that multiple levels of clustering exist within the data, and to deal with this situation *hierarchical* clustering algorithms have also been developed. These algorithms have many natural applications in the sciences as hierarchical classes or groupings are commonly found. Their basis lies in the the use of threshold values for clustering of data based on some form of distance metric. By changing the distance threshold the cluster membership at the new level in the hierarchy is defined.

In the above discussion the issue of unknown numbers of clusters has not been addressed. In this situation, both the cluster features and the number of clusters are unknown. With little information to inform the problem it is difficult to proceed without recourse to prior assumptions concerning the nature of the components or clusters. One approach is to apply algorithms for a range of c values and examine a comparison criterion (a type of profile inference). Alternatively, the whole space of the joint distribution of number and cluster feature must be examined.

In the next section we compare and contrast the forms of spatial clustering which have been developed and attempt to relate these to conventional non-spatial approaches.

1.2.2 Spatial Clustering

Basic random effect modelling in spatial applications considers groupings as random disturbances described by model factors. These factors are often treated as random effects and modelled via prior distributions. More specifically, let us define an arbitrary spatial location as x and a spatial process at that location as $z(x)$. The observed process at x, y say, consists of $z(x)$ and some form of uncorrelated measurement error $\varepsilon(x)$. For example, an additive form is sometimes assumed so that

$$y = z(x) + \varepsilon(x).$$

The spatial process $z(x)$ can include long range effects (e.g. spatial trend) and short range covariance effects. This model is that underlying geostatistical modelling of spatial processes, and kriging is a method which yields point estimates of values of $z(x)$ at various locations (e.g. Chapter 3, Cressie 1993). Often a zero-mean Gaussian process is assumed to underlie the $z(x)$ process. In space-time this model could be simply extended to

$$y = z(\boldsymbol{x}, t) + \varepsilon(\boldsymbol{x}, t)$$

where t denotes a time point.

In these models concentrations of values of the random process are represented by peaks in the resulting fitted surface. These peaks could be regarded as containing location information about elevated values in the data. Of course these are peaks of a process and not concentrations of data locations (as in conventional clustering). However, there is a close correspondence between these ideas. For example if the *density* or *intensity* of points or objects on a map were to be modelled then peaks in the density/intensity surface would hold information concerning clustering and cluster locations. In fact the treatment of the intensity of an object process as a random field is one fruitful approach to modelling clustering of objects in the spatial or spatio-temporal domain. For example, in spatial epidemiology problems, it is possible to model peaks in relative disease risk surfaces and these peaks correspond to local aggregations of cases of disease.

Random effect models in imaging have a structure similar to geostatistical models except they usually use simplified covariance structures to describe spatial correlation (i.e. neighbourhood effects).

Although it is possible to attempt to adapt random effect models to provide information about cluster related features, basic random effect models are mainly designed to provide smooth reconstructions of surfaces. As aggregations or peaks in density are by nature not smooth, these models may not be successful in describing clustering. However, the residual from a smoothed surface estimate could contain cluster information.

In the spatial domain the closest correspondence with conventional cluster analysis can be found in the analysis of point or object processes and derived processes such as counting measures in disjoint regions. Define a spatial study window as T and within this window a realisation of a spatial point process is observed. Denote the n points of this realisation as $\{\boldsymbol{x}_i\}$ $i = 1, ..., n$. It is believed that the spatial point process is clustered. At this point, conventional clustering methods could be used in the reconstruction of the clusters within the pattern, e.g. simple Bayesian classifiers could be used to make a flat classification or more sophisticated mixture models could be applied.

At this point a difference arises in the approaches depending on whether the nature of the underlying spatial point process is to be taken into consideration. Probability models for clustering can be specified in the form of mixtures which do not directly use point process properties. For example, we can assume that the c components of a mixture have densities $f_j(\boldsymbol{x}_i | \theta_j)$ and, in a likelihood formulation, two forms can be examined: a) a classification likelihood

$$L_{class} = \prod_{i=1}^{n} f_{\gamma_i}(\boldsymbol{x}_i|\theta_{\gamma_i})$$

where the γ_i are discrete label values: $\gamma_i = j$ if $\boldsymbol{x}_i$ belongs to the jth component (cluster), and b) a mixture likelihood

$$L_m = \prod_{i=1}^{n} \sum_{j=1}^{m} p_j f_{\gamma_i}(\boldsymbol{x}_i|\theta_j)$$

where p_j is the probability that an observation belongs to the jth component. Fraley and Raftery (1998) (see also Banfield and Raftery 1993) use this approach as a basis for a variety of clustering algorithms. Usually the form of the component densities is multivariate normal. The covariance structure of the components determines the form of the clusters likely to be found in these models. A variety of algorithms have been proposed for the estimation of the parameters in these models. The EM algorithm can be used, and is often used when incomplete data are available. In clustering, the complete data would consist of both the $\{\boldsymbol{x}_i\}$ set and a set of labels associating the data to clusters. Often these labels are unknown and hence the $\{\boldsymbol{x}_i\}$ themselves constitute the incomplete data. A complete data likelihood is then formed and the EM algorithm iterates between estimation of the labels and maximisation of the complete data likelihood. A full Bayesian approach to mixture models can be adopted. Binder (1978) outlined many of the issues in this approach – see also Richardson and Green (1997). While these mixture modelling approaches can be and are applied to spatial problems, the basic form of the mixture is quite general and these models are closest to those of conventional cluster analysis. They do not directly employ spatial stochastic process models for the distribution of the events in the clustered pattern.

Models for spatial clustering can also be based on the stochastic process assumed to underlie the spatial distribution of the data. For example, if the $\{\boldsymbol{x}_i\}$ are a realisation of a spatial point process, then it is possible to include features of the process in the model formulation. In the case of clustered point processes a simple model might be the Poisson cluster process, where the points are thought to cluster around unobserved cluster centres. In this approach the cluster features are the centre locations and a variance parameter determining the cluster spread. This approach is also based on a mixture of cluster components where the centres form a hidden point process. In this approach the approximate first order intensity of the process is defined (conditional on the centre process) to be:

$$\lambda(\boldsymbol{x}|\boldsymbol{c}_j) = \rho \, f \left\{ \sum_{j=1}^{m} h(\boldsymbol{x} - \boldsymbol{c}_j; \boldsymbol{\theta}) \right\}$$

where ρ is the overall rate of the point process, f is a link function, h is a cluster distribution function and c_j are now the centre locations. The h function can take a wide variety of forms depending on the assumed form of the clusters of interest and is not required to be a density. The θ vector now contains only non-locational parameters.

A likelihood can be formed and a full Bayesian analysis can proceed by introducing prior distributions for the number of centres, their locations and other parameters. Special MCMC algorithms have been introduced to sample the posterior distribution of number and location of centres (Birth-Death-MCMC) (Geyer and Møller 1994, Geyer 1999, Stephens 2000). This approach can be generalised or modified in a number of ways.

First, the hidden cluster object process can take different forms: for example linear features could form the centres and so a hidden line process may be assumed. Second, the rigid formulation of parameterisation can be relaxed by allowing the cluster distribution function to be estimated nonparametrically, so that the parameter vector θ includes smoothing constants. These types of models can be extended to model applications where environmental heterogeneity is present, as with modulating functions in spatial epidemiology (Lawson 2000) or to distinguish multi-type processes in geosciences (Cressie and Lawson 2000). Clustering of non-point objects can also be considered in a similar way.

Difficulties arise in all the above approaches when clustering cannot be simply parameterised by a simple set of components or hidden process. For example, clusters of events could form around linear features or have a spatially varying form which cannot be simply modelled by such components. This tends to suggest that a more fully nonparametric approach to clustering should be pursued. The idea of using nonparametric smoothing to isolate residual areas of excess intensity in a spatial window has been pursued by a number of workers (e.g. Kelsall and Diggle 1998, Godtliebsen et al. 2001a). Another approach is to assume a different form of point process model where the intensity of the process is governed by a spatial process which does not necessarily parameterise the clustering tendency but rather models the random variation. Log Gaussian Cox process models fall into this class as described in Chapter 3 (see also Møller et al. 1998). Some more recent developments in clustering focus on models which consist of partitions of the intensity space. These models provide a flexible locally varying alternative to global models for the process intensity (Denison and Holmes 2001, Denison et al. 2002a).

Hierarchical clustering can be achieved in the spatial domain using standard methods or the model-based mixture methods described above. In the spatial domain, it is natural to consider the equivalence of hierarchical clustering with changes in the spatial scale of clustering. Hence methods which allow there to be varying scales in the clustered pattern are equivalent in this sense. The hidden point process methods do allow multiscale phenomena to be examined as they sample the range of cluster number (m) and cluster spread or variance. For a given realisation of points, these are correlated and so large scale clustering is likely to be sampled when small values of m are visited. In general one of the advantages of the use of MCMC algorithms for sampling the joint distribution of the number and cluster feature vector, within a Bayesian setting, is the ability to examine non-flat classifications.

The modelling of clustering in counting processes is closely related to that of point processes in the spatial domain. Where counts of point data are observed in disjoint subregions of a window it is possible to apply many of the above algorithms. Often the conditional expectation of the subregion count is modelled as an integral of the underlying event intensity. Hidden processes can still be estimated in this way, although the spatial information is now at an aggregated level and hence less differentiated (Lawson 1997). Count data arises frequently in spatial epidemiology as well as in spatial ecology. In the former case, there have been a few recent attempts to apply partition models to counts (Knorr-Held and Rasser 2000) as well as using special image priors for count clustering (Gangnon and Clayton 2000).

Finally it can be noted that the field of spatial and spatio-temporal cluster modelling is currently expanding considerably and is likely to be a field for much fruitful research during the new millennium.

1.3 Notation and Model Development

In this section we introduce some of the relevant concepts which underpin the topics covered in depth within this volume. Our intention is to provide a brief account of the basic issues and the notation used in later chapters.

To this end, we will consider the topics as divided into spatial and spatio-temporal models and applications, and within these sections we will consider a broad division between models designed for random variation (often via random effects) and those models designed specifically to detect or recover random objects (such as cluster locations). In addition, we will introduce the concepts associated with other localised models for random variation (such as partition modelling).

In the spatial domain it is useful to consider both parametric and non-parametric approaches to cluster analysis. Before considering parametric approaches, we introduce the basic ideas associated with nonparametric analysis. It should be borne in mind that many nonparametric approaches in the spatial domain can be easily extended to the spatio-temporal domain or to other higher dimensions. In Chapter 2, an approach to nonparametric identification of clusters in the spatial domain is described, and in addition, the extension to higher dimensions is also explicitly described in that work. In the next section we introduce ideas underpinning the approaches there described.

1.3.1 Nonparametric Approaches

It is often the case that only a vague definition of clustering, or what constitutes a cluster, is available. This may be the case at an exploratory stage of analysis, or it may be that the clusters examined can take a wide, or perhaps ill-defined, variety of forms. Often, in this case, areas of a study region which display *unusual* aggregations of items/events are of interest. The definition of *unusual* being made with reference to some null distribution of aggregation. In spatial epidemiology, the null distribution would be the distribution expected from the population at risk of the disease of interest. In other applications, the null distribution could be *complete spatial randomness* (CSR) and departures from such a distribution are to be detected and possibly interpreted as clusters.

Often in this broad definition, clusters are regarded as areas of elevated aggregation where the degree of aggregation surpasses some threshold. Often this threshold is defined by a critical p-value, and any areas on the resulting map which exceed this critical value can be regarded as valid clusters. Usually only areas of elevated aggregation are of interest, although it is possible that areas of particularly low aggregation may also be of interest.

The group of methods known as *bump-hunting* describe this area of nonparametric clustering (for example, see Section 9.2 Scott 1992). These methods rely on the use of nonparametric density estimation methods, such as kernel density estimation, to estimate the local density surface of events on the map. Subsequently, the peaks of this surface (*bumps*) are detected and a threshold value is applied to allow a significance decision. Bump-hunting, in the spatial domain, consists of a set of methods which are designed to detect peaks in a nonparametric density estimate surface. In some special applications, such as point processes, the first order intensity of the process is to be estimated and special edge-corrected estimators are available (Diggle 1983, Diggle 1985a). In other cases, ratios of intensities must be estimated such as in spatial epidemiology (Bithell 1990, Lawson and Williams 1993, Kelsall and Diggle 1995b). Here we discuss density estimation and bump-hunting.

Density/Intensity Estimation For our purposes, the complete set of locations of events/items within a study window is defined to be a *realisation* of an event process. The density of events within the study window will be estimated via nonparametric density estimation. Define the realisation of events as $\{x\}_{i=1,...,n}$. A bivariate (product) kernel estimator of the density, $f(x)$, at arbitrary location $\{s, u\}$, $\widehat{f}(s, u)$ say, can be specified as:

$$\widehat{f}(s, u) = \frac{1}{nh_1h_2} \sum_{i=1}^{n} \left\{ K\left(\frac{s - x_{1i}}{h_1}\right) \times K\left(\frac{u - x_{2i}}{h_2}\right) \right\}. \quad (1.2)$$

Here the spatial (x,y) coordinates of the event locations are specified as

$\{x_{1i}, x_{2i}\}$, and the kernel function $K(.)$ is defined as a univariate density function centred on zero. A commonly used form of kernel is the Gaussian which places symmetric normal form on the contribution of the ith data point to the local density. In this approach the local density of events is controlled by smoothing constants h_1 and $h_{2.}$, one for each direction. If no separate directional smoothing is desired, then a common smoothing parameter can be applied in each direction. The smoothness of the resulting density surface depends crucially on the value of the smoothing constant(s) (*bandwidths*). A variety of methods have been developed to provide objective criteria for bandwidth selection, such as cross-validation (Section 6.5 Scott 1992).

In applications in spatial clustering, any smoothing of the spatial variation in aggregation could potentially reduce the evidence for clustering. If the first order intensity of a point or event process is locally elevated, then this may provide evidence for a local cluster of events. Over-smoothing of the density/intensity might reduce the elevation of the density or first order intensity locally, and so reduce evidence for clustering. Hence over-smoothing by improper bandwidth choice could be a concern here. Of course, examining the evidence for clustering at *different* bandwidths may shed light on the persistence of clusters (and perhaps their significance) and the pattern of clustering at different spatial scales. In Chapter 2, this idea is exploited via graphical tools to find features which arise at different scales.

Bump-Hunting   The basic focus of bump-hunting in the spatial domain is the task of finding modes or peaks of a spatial distribution of events/features. There are two distinct approaches to this task. The first approach regards areas of elevated aggregation as evidence of a localised peak. With more extreme aggregation the peaks become more elevated and so this approach seeks to determine the areas of the local density/intensity which have 'significantly' elevated levels. In this approach there is an implicit assumption of a background, or null, variation in risk under a suitable null hypothesis. This approach has been advocated for the isolation of areas of excess relative risk in spatial epidemiology, and is a potentially useful tool for exploratory analysis of excess variation in applications where background (null) spatial variation or heterogeneity is to be included. In the example of spatial epidemiology, an intensity ratio is formed from the intensity estimates of disease case events on a disease map, and the intensity estimate of control disease events on the same map. The control disease events are used to represent the null or background variation. Let $\widehat{f}(\boldsymbol{x})$ be the intensity of cases at arbitrary location $\boldsymbol{x}$, and $\widehat{g}(\boldsymbol{x})$ the intensity of control events. Then the ratio

$$\widehat{R}(\boldsymbol{x}) = \frac{\widehat{f}(\boldsymbol{x})}{\widehat{g}(\boldsymbol{x})} \tag{1.3}$$

will display excess risk over the study area. Further, it is possible to construct a Monte Carlo p-value surface by simulating case event realisations from the null intensity $\widehat{g}(\boldsymbol{x})$. The rank of the pointwise $\widehat{R}(\boldsymbol{x})$ compared to the ratios computed from the $\widehat{f}(\boldsymbol{x})$ for the simulated realisations, gives a pointwise significance surface. Details of this procedure are given in Kelsall and Diggle (1995b) and Chapter 6 of Lawson (2001).

While the excess variation procedure is a useful exploratory tool it does not use all the information available from the surface concerning the local peaks of variation. Much of the literature in bump-hunting for density estimation has focussed on the estimation of derivatives of density estimates. The derivative shows gradient information and can point to areas of elevated variation. The kernel estimate of the derivative for the univariate case is a sum of the derivative of the kernel function. In the spatial case directional derivatives must be specified, and similar results arise. By examining the nature of a map of first and second directional derivatives it is possible to assess the degree of peaking on the map. The SiZer and S^3 procedures described in Chapter 2 address these gradient issues for one and two dimensions.

The final issue related to bump hunting is the use of scale information in the construction of thresholds for accepting areas as clusters. The scale of clusters relates closely to the number of clusters and so these also relate to the degree of smoothing and hence the bandwidth used. The notion of a critical bandwidth beyond which clusters appear/disappear has appeal, and a test was developed in the univariate case by Silverman (Section 9.2 Scott 1992). This idea is further examined for the spatial (and higher dimensional) case in Chapter 2 where the idea of 'significance in scale-space' is introduced.

1.3.2 Point or Object Process Modelling

The parametric modelling of the *locations* of point or object events can be approached in a variety of ways. Unlike the nonparametric approaches described in 1.3.1, models for the location and nature of clusters must be introduced and the appropriate inferential machinery developed. Usually the available data are a realisation of events (points or objects) within a study window, and the spatial distribution of the events is assumed to be governed by a parametric model. Models from within stochastic geometry are often used to describe the spatial distribution. In the case of clustering, the Neyman-Scott process is often assumed where a notional set of unobserved cluster centres is thought to explain the clustering of the observed data. The development of such models and their estimation are discussed in detail in chapters 3 and 4 in this volume. Another approach is to assume that the local aggregation of events (first order intensity) has different levels or partitions (see Chapter 6). The use of Markov chain Monte Carlo and perfect sampling are also featured in Chapters 4, 5, 6 and 14. Models where the distribution of events is related to a hidden process of centres constitute one form of Cox Process, whereas log Gaussian Cox processes, as described in Chapter 3, have a governing intensity which is the realisation of a Gaussian spatial process. In the first case, the locations of centres are explicitly modelled, whereas in the second case areas of elevated intensity (peaks in the Gaussian process) correspond to local aggregations.

While the focus of modelling of the spatial distribution of events is a primary concern in clustering, it is also the case that events may have associated measurements which also require to be modelled. For example, the location of trees in a forest may be of interest but the height or width of the trees may have also been measured and may possibly relate to the locational preferences of the trees. Another example, from spatial epidemiology, would be the diagnostic history of cases of disease located at residential addresses. In each case, the location of an event is *marked* with an associated measurement. This implies that models for locations and for marks must be considered. It may be that the marks themselves are clustered spatially and/or the locations are clustered. One approach to this problem is to model the marks conditionally on the locations (as if they were a fixed sampling network). Of course this does not explicitly allow for interaction between the two processes, but it does demonstrate the close linkage between models for random processes (marks) and models for locations (events).

1.3.3 Random Effect Modelling

Parametric modelling of spatial data is often formulated in terms of traditional fixed and random effect components. Usually a random variable is observed at spatial locations, and this is regarded as the observation of a spatial random process at these locations. The observed values of the spatial process are $\{y_i\}_{i=1}^n$ for n locations. The true process at arbitrary location x is defined to be $z(x)$. The model relating these components is defined to be :

$$y|z \sim g\{z(x); \theta\}, \tag{1.4}$$

where $g\{.\}$ is defined as a probability distribution describing the variation around the true model, and θ represents a set of parameters.

The true process, $z(x)$, can be composed of a variety of effects. If a Gaussian measurement error model is assumed, then

$$y|z \sim N\{z(x), \sigma^2\}.$$

The error term (ϵ) is usually assumed to have zero mean and be spatially uncorrelated, and so $y = z(x) + \epsilon$. In the simple case where variation in $\{y_i\}$ is thought to be related to a set of explanatory fixed covariates, possibly measured at the n sites, then we can define a generalised linear or additive regression model. Here we introduce a design matrix F which consists of p covariates and is of dimension $n \times p$ and α, a $p \times 1$ vector of parameters. Note that the covariates can be spatially dependent or not as the context dictates. For example, if there is no spatial correlation inherent in the process, then a simple generalised linear model might be appropriate for which

$$y|z \sim g\{f(F\alpha), \theta\}$$

where $f(.)$ is a suitable link function. In the simple normal model case then this is just:

$$y|z \sim N\{f(F\alpha), \sigma^2\}$$

In the case of additive models (GAMs) then $f(F\alpha)$ will be partitioned into different smoothed components.

In the general case (1.4), $z(x)$ could be extended to include random effects as well as fixed effects. For example, $z(x) = \mu + z_1(x) + z_2(x)$, where the fixed effect $\mu = F\alpha$ could be specified. Here, the extra components $(z_1(x), z_2(x))$ can represent effects which are not observed but may yield extra variation in the observed value of y. In spatial epidemiology, these effects could represent *unobserved* covariates such as deprivation, lifestyle, age, gender or ethnic origin or even unknown effects as yet not measured, or even unmeasurable. In image analysis this extra variation may simply be attributable to unknown unobserved effects. Generalised linear mixed models (GLMMs) are special cases of these models (Breslow and Clayton 1993).

The two extra components $(z_1(x), z_2(x))$ are often specified to model spatially correlated or uncorrelated heterogeneity. In this way any extra peakedness in the true or observed process will be included in the model. In many situations, these *heterogeneity* effects are also termed *clustering* effects. However, while these effects tend to mimic the effect of peaks in, say, a clustered event process, the peakedness does not relate to the aggregation of an underlying event process, but simply to extra variation in the observed process or underlying true process. Of course, if the intensity of an event process was a Gaussian or log Gaussian spatial process then peaks in the Gaussian process would relate to areas of elevated aggregation in the events (for more details see the discussion of Cox processes in Chapters 3, 4, 5 in this volume).

Often the modelling of $z_1(x)$, $z_2(x)$ is easily carried out via Bayesian hierarchical modelling, although a wide range of alternative approaches could also be adopted. In the Bayesian approach the processes $z_1(x)$, and $z_2(x)$ are given prior distributions which define the spatially structured (correlated) term $z_1(x)$ say, and the overdispersion term $z_2(x)$ say. One version of this specifies that the $z_1(x)$ has a spatial Gaussian process prior with specified covariance model. This approach can be described as generalised linear geostatistics (Diggle et al. 1998). In Chapter 11 in this volume an extension of this approach is described where a discrete data Poisson model is assumed. Another approach, popular in applications such as spatial epidemiology, is to assume a intrinsic Gaussian prior distribution for the spatially structured component (Besag et al. 1991), where dependence neighbourhoods are specified instead of a parametric covariance model. The uncorrelated component is often assumed to have a zero-mean Gaussian prior distribution. In chapter 9 a new class of skew-Gaussian models for geostatiscial prediction are discussed within a Bayesian hierarchical framework.

More traditional non-Bayesian approaches include universal kriging where the value of y is thought to be modelled by a linear predictor and the second order properties of the true underlying process are described by a covariance model. In general, we should note that random effect modelling does not have as its primary focus the estimation of the location of clusters or areas of unusually elevated variation.

1.3.4 Partition Modelling

An alternative approach to modelling of spatially localised variation is to assume that the spatial process considered is underlain by a mixture (sum) of components and these components are to be estimated. A review of such models is given in Chapter 7. The basis of these models is similar to the random effect models of Section 1.3.3 in that the observed data at the ith sample location is modelled by

$$y_i = \mu + z(\boldsymbol{x}_i) + \epsilon_i,$$

but here the spatial process is thought to consist of a sum of spatial components

$$z(\boldsymbol{x}) = \sum_{j=1}^{J} w_j z_j(\boldsymbol{x}|\boldsymbol{c}_j) \tag{1.5}$$

where the $\{w_j\}$ are weights and $\{\boldsymbol{c}_j\}$ are the centres of a tiling of the study area. The extension to this for non-Gaussian non-centred data would be of the form

$$y|z \sim g\{z(\boldsymbol{x}); \theta\},$$

where $z(\boldsymbol{x})$ is defined as in equation (1.5). In Chapter 8 a different formulation seeks to model discrete count data clusters as partitions where clustering is controlled by a prior distribution which penalises certain forms of configuration.

1.3.5 Spatio-Temporal Process Modelling

Extensions to the spatial modelling described in Section 1.3.3 can be made to the spatio-temporal case. In this extension the data observed at a specific location in space-time is to be modelled. The type of data arising depends on the measurement units adopted and can vary considerably depending on the application context. For example, in disease mapping applications it may be possible to obtain the residential address and the date of diagnosis of a case. In this case a complete realisation of a spatio-temporal point process may be observed. However in other cases, only counts of events within fixed spatial or temporal periods are available, or measurements at spatial monitoring sites are made at fixed time intervals. For fixed spatial sites indexed as $i = 1, ..., n$ and temporal periods t_j ($j = 1, ..., m$) then a model for the observation y_{ij} could be specified in general as:

$$y_{ij}|z(\boldsymbol{x}_i, t_j) \sim g\left\{z(\boldsymbol{x}_i, t_j); \theta\right\},$$

where the components of the model will depend on context. In Chapter 12 a decomposition of $z(\boldsymbol{x}_i, t_j)$ into spatial and temporal basis functions (principal kriging functions) is examined as well as other possible models. The models including correlation are based on parameterised covariance functions. Other forms can be specified for the structure of $z(\boldsymbol{x}_i, t_j)$. In epidemiology, a number of models have been proposed for space-time counts of disease (Knorr-Held and Besag 1998, Cressie and Mugglin 2000, Waller et al. 1997b, Bernardinelli et al. 1995, Sun et al. 2000, Zia et al. 1997, Knorr-Held 2000). These often include separate forms of spatial and temporal variation including spatial and temporal correlation and spatio-temporal interaction terms. These models seek to describe the variation in disease risk across a study region but do not focus on cluster location *per se*.

Extensions to partition models in space-time can also be made. In Chapter 13 an application of spatio-temporal partitioning to an imaging problem is described. In their model, brain responses are assigned via a mixture model to a finite number of levels.

The spatio-temporal extension for event or point process clustering can also be specified in a straightforward manner. The point process is governed by an intensity of form $\lambda(\boldsymbol{x}, t) = f\{z(\boldsymbol{x}, t), \theta\}$ which is specified to depend on the realisation of a spatio-temporal process. This process could depend on the (unobserved) locations of cluster centres (a hidden cluster centre process) or could be designed to be dependent on the proximity of observed events up to a particular time. Alternative formulations where $\log\{\lambda(\boldsymbol{x}, t)\}$ depends on a spatio-temporal Gaussian process (a log Gaussian Cox process) could also be specified. However they do not directly model the locations of clusters in space-time. The division between data-dependent and hidden process dependent models for clusters is discussed in more detail in Chapter 14. In that chapter the connection between point processes and binned counts is examined and a hidden process model is applied to a binned realisation of events.

PART I

Point process cluster modelling

CHAPTER 2

Significance in Scale-Space for Clustering

F. Godtliebsen   *J.S. Marron*   *S.M. Pizer*

2.1 Introduction

An intuitive, visual approach to finding clusters in low dimensions is through the study of smoothed histograms, e.g. kernel density estimates. Scale-space provides a useful framework for understanding data smoothing. See Lindeberg (1994) and ter Haar Romeny (2001) for an excellent overview of the extensive scale-space literature.

The scale-space approach has allowed practical resolution of several long-standing problems in the statistical smoothing literature. See Chaudhuri and Marron (1999, 2000) for detailed discussion. For example, the classical problem of choice of the level of smoothing (bandwidth) can be viewed in an entirely new way using scale-space ideas. In particular, instead of choosing one level of smoothing, one should consider the full range of smooths (the whole scale-space). This corresponds to viewing the data at a number of different levels of resolution, each of which may contain useful information.

For clustering purposes, this simultaneous viewing of several different levels of smoothing incurs an added cost of interpretation. In particular, it becomes more challenging to decide which of the many clusters that are found at different levels represent important underlying structure, and which are insignificant sampling artifacts. An overview of some solutions to this problem is given in Section 2.2. These solutions involve scale-space views of the data (i.e. a family of smooths), which are enhanced by visual devices that reflect the statistical significance of the clusters that are present.

In keeping with the visual nature of these new methods, only one and two dimensional cases are presented. Certainly higher dimensional clustering is of keen interest, but visual implementation in higher dimensions represents a very significant hurdle. For now, dimension reduction methods need to be applied first, before these approaches can be used in higher dimensions.

In Section 2.3 we propose a new enhancement of the two dimensional version, based on the natural idea of contour lines.

Finally there is some discussion of interesting future research directions in Section 2.4.

2.2 Overview

There are a number of different approaches to the nonparametric assessment of the statistical significance of clusters in one and two dimensions. This problem was called "bump hunting" by Good and Gaskins (1980). A wide range of approaches to this topic may be found in the papers Silverman (1981), Hartigan and Hartigan (1985), Donoho (1988), Izenman and Sommer (1988), Müller and Sawitzki (1991), Hartigan and Mohanty (1992), Minnotte and Scott (1993), Fisher et al. (1994), Cheng and Hall (1997), Minnotte (1997) and Fisher and Marron (2001). Many of these approaches are only concerned with the number of significant clusters.

In this chapter we discuss more visual approaches to significant clusters, that make more explicit use of scale-space ideas. An advantage of the visual approach is that one also learns where clusters are located. Viewing through scale-space reveals the levels of resolution at which each cluster appears.

Nonparametric statistical inference for clustering in one dimension was developed by Chaudhuri and Marron (1999). Their method is called SiZer, for "SIgnificance of ZERo crossings". SiZer finds clusters through the study of the slope of the smooth histogram. A cluster is significant when the slope of the curve is significantly positive on the left, and significantly negative on the right. In particular, when there is a statistically significant zero crossing of the derivative. An example is given in Figure 2.1.

Flow cytometry studies the presence as well as the percentage of fluorescence marked antibodies on cells. The fluorescence of individual cells is measured, and the results are binned, into bins called "channels". The top of Figure 2.1 shows a bar graph of square root bincounts, from a single experiment. The bar graph suggests that there are two clusters in the data. Are the clusters statistically significant? Or could they be mere artifacts of the sampling process? The issue is not 100% clear, because the peaks contain some bars that dip below the taller bars located in the valley between. SiZer aims to address this issue.

The heights of the bars in the top panel of Figure 2.1 are shown as dots in the middle panel, which also shows the scale-space as the family of curves. If the raw fluorescence levels were available, the scale-space curves would be kernel density estimates. However SiZer also works in terms of counts, using local linear smoothing to obtain the scale-space, which is done here. The inference done by SiZer is shown in the map in the bottom panel of Figure 2.1. The horizontal (x) axis of this map is the same as the x-axis of both of the panels above, i.e. "location". The vertical (y) axis of the map is "scale", i.e. bandwidth of the smooth, on the log scale. Thus each row of the

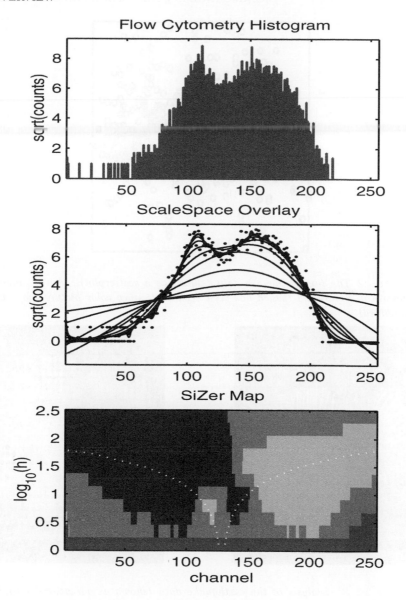

Figure 2.1 *SiZer analysis of the flow cytometry data. Original square root bin counts in the top panel. Family of smooths in the middle panel. SiZer analysis, showing two significant clusters in the bottom panel.*

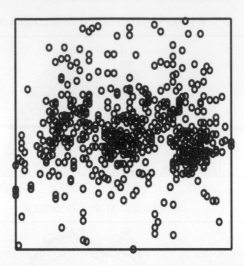

Figure 2.2 *The raw earthquake data (shown as a scatterplot). An S^3 analysis appears in Figure 2.3 (which has been separated, since color pictures have been combined).*

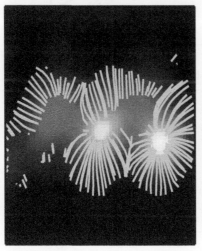 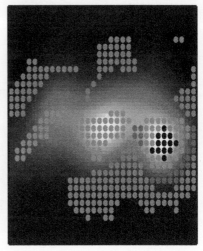

Figure 2.3 *S^3 analysis of the earthquake data (shown as a scatterplot in Figure 2.2), using gradient streamlines (left panel) and curvature dots (right panel). The right cluster is clearly statistically significant, the middle cluster is not quite conclusive, the left cluster is less well defined.*

SiZer map corresponds to one of the curves in the middle panel. The SiZer statistical inference is based on a confidence interval, for slope (derivative) of the smooth at each location, and at each scale. When the confidence interval is completely above 0, the smooth is significantly increasing, and the color blue (shown here as a dark shade of gray to minimize the need for color printing) is used. When the confidence interval is entirely below 0, the smooth significantly decreases, and this location in the map is colored red (shown here as a light shade of gray). In the indeterminate case, when the confidence interval contains 0, the intermediate color of purple (shown here as a the lighter intermediate shade of gray) is used. The fourth SiZer color is gray (the darker intermediate shade here), which is used at locations and scales where there is not enough data in each kernel window for reliable statistical inference. This SiZer map shows that both clusters are statistically significant. In particular the left hand cluster around channel 110 is seen to be "really there" because of the large blue (dark) patch to the left, and the smaller red (light) patch on the right (flagging significant increase, followed by decrease). The same holds for the larger cluster around channel 160.

The SiZer visualization is very useful in one dimension, but does not extend easily to the two dimensional case. One reason is that an overlay view of the family of curves (the scale-space) is no longer possible. A more serious reason is that the SiZer foundation of "significantly sloping up or down" no longer makes sense in two dimensions. Proposals for some completely new visualizations of statistically significant features, called S^3 for "significance in scale-space", were made in the case of two dimensional images by Godtliebsen et al. (2001a). Some closely related proposals for two dimensional smooth histograms, and thus for finding significant clusters, were made by Godtliebsen et al. (2001b). The problem of lack of availability of overlays in two dimensions is addressed by the construction of movies where time is the scale (i.e. level of smoothing). The problem of statistical inference is addressed by adding visual enhancements to the movie. Some of these are illustrated in Figures 2.2-2.3.

Figures 2.2-2.3 provide a visual clustering of the earthquake data from Section 4.2 in Wand and Jones (1995), shown as a scatterplot in the top panel. These data record the locations of epicenters, in longitude (the x coordinate, with 122 subtracted for numerical convenience) and latitude (the y coordinate, with 46 subtracted) of earthquakes in the Mount St. Helens area of the United States.

The left hand panel of Figure 2.3 demonstrates the streamline version of S^3. The green "streamlines" are the visual cues indicating statistically significant structure. These are based on the gradient of the gray level surface, which is the direction of maximal change. These green curves essentially show the direction that a drop of water would follow as it moves down the surface. However, lines are only drawn in regions where the gradient is statistically significantly different from 0, i.e. where there is a significant slope.

The gray level plot, together with the streamlines, suggest three clusters, as does the scatterplot on the top. The streamlines show that there is strong evidence only for the right cluster being statistically significant, as indicated by the ring of streamlines pointing towards the peak, that are completely around this cluster. The middle cluster has streamlines pointing towards the peak most of the way around, which is a suggestion of a cluster, but not conclusive statistical evidence. The left cluster is much less convincing (in the sense of statistical significance), because there are few streamlines pointing towards its peak. Of course it must be kept in mind that statistical significance is necessarily one sided. Streamlines give strong evidence of presence of a feature, but lack of streamlines only indicates the evidence is not strong enough to be sure, and does not prove absence of a cluster.

To save space, only one scale, i.e. bandwidth, is shown in Figure 2.3. This is $h = 4$, which was chosen for presentation purposes, after viewing the full scale-space. For data analytic purposes, this viewing of the full scale-space is essential, and we suggest doing this as a movie. We recommend viewing the movie version of the left side of Figure 2.3, in the file SSScntr1Fig2a.avi in the web directory http://www.unc.edu/depts/statistics/postscript/papers/marron/SSS_cntr/. The movie format is AVI, which is easily viewable on most computers without the need of downloading an extra viewer.

The right panel of Figure 2.3 shows an alternate version of S^3. This time the statistical inference is based on curvature. Curvature is conveniently described in two dimensions using the eigenvalues of the Hessian matrix. At a grid of image locations, statistical significance of these eigenvalues is assessed. Colored dots, overlaid on top of the gray level image, provide quick visual access to this information. The following table indicates colors that are used for the various curvature cases, depending on the largest eigenvalue $\widehat{\lambda}_+$, the smallest eigenvalue $\widehat{\lambda}_-$, and the appropriate quantile $\widehat{q}_{\widehat{T}}$. See Godtliebsen et al. (2001a) for details of the derivation, including the choice of $\widehat{q}_{\widehat{T}}$. The latter requires substantial effort, even when the data are exactly Gaussian, because the joint distribution of the eigenvalues $\widehat{\lambda}_+$ and $\widehat{\lambda}_-$ is non-standard. Appropriate rescaling and tabulation of this distribution using simulation methods is done in Godtliebsen et al. (2001a).

color	feature	characterization
yellow	hole	$\widehat{\lambda}_+, \widehat{\lambda}_- > \widehat{q}_{\widehat{T}}$
orange	long valley	$\widehat{\lambda}_+ > \widehat{q}_{\widehat{T}}, \left\|\widehat{\lambda}_-\right\| < \widehat{q}_{\widehat{T}}$
red	saddle point	$\widehat{\lambda}_+ > \widehat{q}_{\widehat{T}}, \widehat{\lambda}_- < -\widehat{q}_{\widehat{T}}$
purple	long ridge	$\left\|\widehat{\lambda}_+\right\| < \widehat{q}_{\widehat{T}}, \widehat{\lambda}_- < -\widehat{q}_{\widehat{T}}$
dark blue	peak	$\widehat{\lambda}_+, \widehat{\lambda}_- < -\widehat{q}_{\widehat{T}}$

Most of these colors appear in the right side of Figure 2.3. Again only a single scale has been selected, $h = 5.19$, after viewing the full scale-space. This scale is larger than that chosen for the streamline analysis in the left panel, because curvature estimates feel noise more strongly than slope estimates, so more smoothing is needed for similar inference. Viewing the full scale-space is again recommended, using the movie file SSScntr1Fig2b.avi in the above web directory. As in the above analysis, the right cluster in the data comes through very strongly. In particular there are a number of dark blue dots. The center cluster is less clear, showing all purple dots, which only show existence of a ridge, not a cluster. Some suggestion that the middle cluster is separate is provided by the red saddle point dots between the clusters, but this is not conclusive, because these also appear on a ridge that then slopes upwards. The potential third cluster on the side shows up less strongly here than in the streamline analysis, because at this coarser scale (needed for adequate noise reduction) it is nearly smoothed away. The orange dots highlight locations where "clusters emerge from regions of no data".

A disappointing aspect of the analysis of Figure 2.3 is that only the right hand cluster is statistically significant in this sense. A possible approach to investigating the significance of the other clusters is to adjust the statistical inference, via the level of significance, α. In Figure 2.3, and all other examples in this chapter, the standard $\alpha = 0.05$ is used. Results for the less stringent case of $\alpha = 0.2$ are viewable in the movie files SSScntr1Fig2c.avi and SSScntr1Fig2d.avi. These are not shown here to save space. The most interesting result was that at the scale $h = 6.17$, at least one dark blue dot appeared at the top of each of the three clusters, providing evidence that all 3 clusters were significant (at this lower level).

All smoothing methods used in the chapter are kernel based methods. For more background information on these, see for example Scott (1992), Wand and Jones (1995) and Fan and Gijbels (1996). The Gaussian kernel function is used everywhere in this chapter, as this has the most appealing scale-space properties, see Lindeberg (1994).

An important technical component of these methods is correct simultaneous statistical inference. In particular, many inferences are performed here at one time. In such situations, naive implementations of hypothesis tests will result in a number of false positives. Care has been taken to avoid this problem in the construction of SiZer and S^3. See Chaudhuri and Marron (1999) and Godtliebsen et al. (2001a, 2001b) for details.

2.3 New Method

The streamline approach to S^3, shown in the left panel of Figure 2.3, highlights "statistically significant gradient directions" in two dimensions quite well. However, gradients alone are not a particularly intuitive vehicle for

Figure 2.4 S^3 *analysis of the asymmetric volcano simulated example. Streamline only (left panel) and streamline and contour (right panel) versions. This shows that the addition of contours enhances the interpretability.*

understanding "surface shape". This is shown in the left hand panel of Figure 2.4, which is based on a simulated data set, composed of a surface with additive i.i.d. Gaussian noise. The underlying surface is an "asymmetric volcano", that is formed from the volume of revolution of a Gaussian probability density, with mean between the center and edge, with an off-center (towards the right) cylinder lowered to height 0 near the middle. One frame of the scale-space is shown, and it is recommended that others be viewed in the movie `SSScntr1Fig3a.avi` from the above web directory.

After careful contemplation, using visual clues from the underlying gray-level image, it becomes clear that the central streamlines start in the low black region near the center and climb the inner cone in a radial fashion. When they reach the circular crest, they then follow the gradient of the crest towards the left. Finally at the top of the crest (on the left side) the gradient is no longer significant, and the streamlines stop. Streamlines coming from the outer edge climb the outer cone in a radial way, and join the inner streamlines at the crest.

While the streamlines describe the statistically significant aspects of the shape of the surface, substantial thought is required for complete understanding. The goal of this section is to provide additional visual clues, which assist this process, in particular by the addition of contour lines to the S^3 graphics. The result of this is shown in the right hand panel of Figure 2.4. Note that the contour lines, shown in purple, are orthogonal to the green gradient lines. Statistical significance of the feature being illustrated (e.g. a cluster), is shown by only drawing contours in regions where the gradient is statistically significant. This notion of statistical significance is particularly well suited to finding clusters. In particular a significant cluster will be a hill of high density (i.e. light gray), that is highlighted by a purple circle surrounding it. The interpretation of the circle is that everywhere around, the slope is significant, which is a useful notion of "cluster".

The contour lines in the right part of Figure 2.4 provide quicker intuitive understanding of the shape of the surface. The central circular contours clearly show the inner cone of the volcano. The banana shaped higher level contours immediately reveal the shape of the light colored curved ridge on the left side.

These new significant contours are constructed using the same statistical inference methods as for the streamline version of S^3. In particular, each pixel location is flagged as having either a significant, or an insignificant gradient. Streamlines are drawn by using a step-wise procedure, following gradient directions, with some random starting values. This methodology could be used for the contours, by simply stepping orthogonally to the gradient. However, contours are not easy to draw in this way, because they generally form closed curves, which are not simple to construct using stepwise procedures involving accumulating numerical errors. But contours have the advantage that many ready made functions to draw them are available. We use the generic contour subroutine in Matlab, with deletion of parts of the contours in regions where the gradient is not significant.

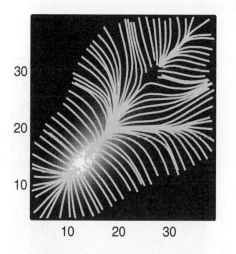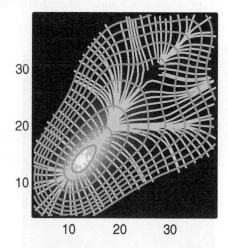

Figure 2.5 *Shows streamline and contour analysis is more effective than streamlines only, for the Melbourne temperature data.*

Figure 2.5 shows some S^3 analyses of the Melbourne temperature data. The raw data here is a lag one scatterplot of daily maximal temperatures in

Melbourne, Australia, over a 10 year span. The x axis shows yesterday's maximum, and the y axis shows today's maximum. A simple weather prediction idea is to use yesterday's maximum to predict today's maximum. If that were exactly correct, all the data points would lie on the diagonal 45° line. Using a conditional density estimation analysis, Hyndman et al. (1996) showed visually that there is a "horizontal ridge of high density", i.e. a ridge where today's maximum is 20 degrees (Celsius). Verification of the statistical significance of this ridge (and evidence for a related vertical ridge) were provided by Godtliebsen et al. (2001b), using an analysis similar to that shown in the left panel of Figure 2.5. Both the horizontal and vertical ridges have a physical explanation, discussed in Godtliebsen et al. (2001b).

The left hand panel shows the streamline only analysis at the scale $h = 4.36$. Again all scales should be viewed, and are available in the movie file SSScntr1Fig4a.avi at that above web address. The horizontal ridge is visible as a coalescence of streamlines. The diagonal ridge does not show up well because near 30 degrees the gradient along the ridge is no longer significant. The vertical ridge does not appear at this scale, but does show up for smaller scales.

The right panel of Figure 2.5 shows that adding contours again enhances the analysis, for the same scale (see also the full scale-space movie in SSScntr1Fig4b.avi), $h = 4.36$. Curves in the contours provide a stronger visual impression of "ridges" than is available from the streamlines. Even a hint at the vertical ridge is available in this way. The contour plot provides much more immediate visual insight about the diagonal and horizontal ridges.

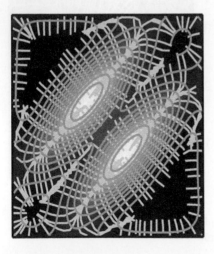 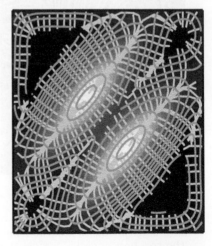

Figure 2.6 *Simulated example showing the difference between equal height and modified quantile contour spacing.*

An issue in the construction of contours is their spacing. Most of the examples in this chapter use equal height spacing. However, this is sometimes inappropriate. This is demonstrated in a simulated example in Figure 2.6. Here the underlying surface is two elongated peaks, as can be seen from the gray level plot. The left panel of Figure 2.6 shows equal height spacing of the contours. Here the scale is $h = 4$, but other scales revealed similar effects (see the movie version in the file `SSScntr1Fig5a.avi` in the same web directory). Note that in the large flat areas on the upper left and the lower right, the contours are somewhat deficient, in the sense that no contour appears for long stretches of the significant green streamlines.

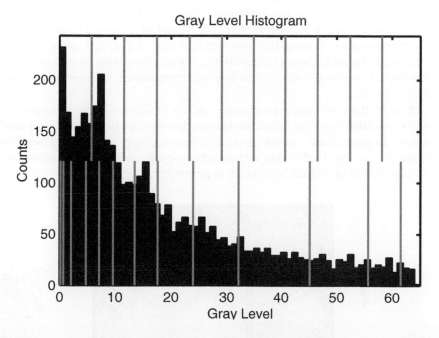

Figure 2.7 *Gray level histogram for image shown in Figure 2.6. This contrasts equal height contour spacing (represented as gray bars in the top half), with modified quantile spacing (represented as gray bars in the bottom half).*

A simple solution to this problem is to add more contours. However, this is not satisfactory, because the contours then become too dense in other regions. A better solution is to use different types of contour spacing. An alternate approach to contour spacings is explained using the gray level histogram shown in Figure 2.7. The gray bars in Figure 2.7 appearing in the top half of the plot show how equally spaced height contours relate to the population of gray levels in the image. Note that a quite large number of pixels near the left edge of the population (nearly black) are represented by only a few contour lines. This explains the poor contour performance in the dark areas of the left panel of Figure 2.6.

Figure 2.7 suggests that this problem can be solved by taking the contour heights according to quantiles of the gray level population, as indicated by the gray bars in the lower half. An initial experiment (see the movie SSScntr1Fig5c.avi in the same web directory) with equally spaced quantiles suggested it was worth including more contours at each end. Some experimentation led us to suggest including 0.1 and 0.4 times the quantile spacing at the lower end, and making a symmetrical inclusion at the upper end. This is done for the gray bars on the bottom of Figure 2.7, and for the contours in the right panel of Figure 2.6, see the movie version in SSScntr1Fig5b.avi in the same web directory. The equally spaced quantile version is quite similar to the right panel of Figure 2.6, except that the two contours nearest the peak are missing. An advantage of quantile spacing is an additional interpretation. When 9 equally spaced quantiles are used, the region between two consecutive contours encloses ten percent of the pixels.

The Melbourne temperature data, shown in Figure 2.5 is a case where equal spacing does not work, so the modified quantile spacing was used there. In general, we suggest choosing between height spacing and modified quantile spacing, by starting with height spacing, and switching when that is unsatisfactory (which is visually apparent).

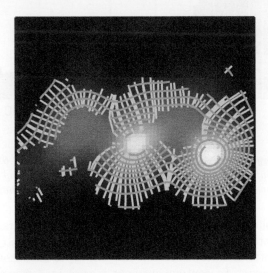

Figure 2.8 *Contour and streamline analysis of the earthquake data.*

In Figure 2.8, the earthquake data from Figure 2.2 are analyzed by adding contours to the streamline analysis from the left side of Figure 2.3 (at the same scale $h = 4$). The full scale-space movie is available in the file SSScntr1Fig7a.avi in the same web directory.

The main lessons about statistical significance of clusters is the same as above, but again the contours make the information more easily accessible.

2.4 Future Directions

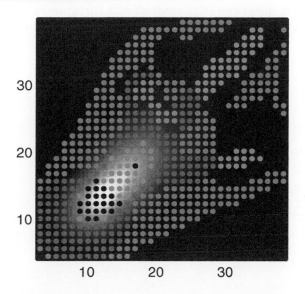

Figure 2.9 *Curvature dots analysis of the Melbourne temperature data.*

While the contour version of S^3 provides an improvement to earlier versions, there are many other possible improvements, and interesting directions for further research.

One such area is to combine the information present in the dot version of S^3 with significant contours. A straightforward approach is to color the contours using the same dot colors. This could provide more immediate interpretation of some of the contours, as well as incorporating additional useful information. Figure 2.9 shows the curvature information available for the Melbourne temperature data. Note that this provides a different compelling evidence for the statistical significance of the ridges. The full scale-space movie version is available in SSScntr1Fig8.avi in the same web directory.

A more complicated, but perhaps more useful extension is to visually represent the statistical significance of the curvature of the contour lines themselves. For example with the contour analysis of the Melbourne temperature data shown in Figure 2.5, it appears that a number of the contours are not convex. Is this concavity statistically significant?

Perhaps the most challenging extension of S^3 is from two dimensions to three. As with the extension from SiZer (dimension 1) to S^3 (dimension 2), the main challenge is the visualization. For example, the scale-space is easily presented as an overlay in one dimension, and a movie in two dimensions, but it is less clear how to present a three dimensional scale-space. After this problem is solved, then careful attention needs to be given to which quantities (e.g. gradient and curvature) should have their statistical significance displayed, and how they should be visualized.

CHAPTER 3

Statistical Inference for Cox Processes

J. Møller   R.P. Waagepetersen

3.1 Introduction

This chapter is concerned with statistical inference for a large class of point process models, which were studied in a seminal paper by Cox (1955) under the name doubly stochastic Poisson processes, but are today usually called Cox processes. Much of the literature on Cox processes is concerned with point processes defined on the real line $\mathbb{R}$, but we pay attention to the general d-dimensional Euclidean case of $\mathbb{R}^d$, and in particular to the planar case $d = 2$ (which covers most cases of spatial applications). However, the theory does not make much use of the special properties of $\mathbb{R}^d$, and extensions to other state spaces are rather obvious. We discuss in some detail how various Cox process models can be constructed and simulated, study nonparametric as well as parametric analysis (with a particular emphasis on minimum contrast estimation), and relate the methods to simulated and real datasets of aggregated spatial point patterns. Further material on Cox processes can be found in the references mentioned in the sequel and in Grandell (1976), Diggle (1983), Daley and Vere-Jones (1988), Stoyan et al. (1995), and the references therein.

To explain briefly what is meant by a Cox process, consider the spatial point patterns in Figure 3.1. As demonstrated in Section 3.3, each point pattern in Figure 3.1 is more aggregated than can be expected under a homogeneous Poisson process (the reference model in statistics for spatial point patterns). The aggregation is in fact caused by a realization $z = \{z(\boldsymbol{x}) : \boldsymbol{x} \in \mathbb{R}^2\}$ of an underlying nonnegative spatial process Z, which is shown in gray scale in Figure 3.1. If the conditional distribution of a point process X given $Z = z$ is an inhomogeneous Poisson process with intensity function z, we call X a Cox process with random intensity surface Z (for details, see Section 3.3). Note that points of X are most likely to occur in areas where Z is large, cf. Figure 3.1. In many applications we can think of Z as an underlying "environmental" process. Aggregation in a spatial point process X may indeed be due to other sources, including (i) clustering of the points in X around the points of another point process C, and (ii) interaction between the points in X. For certain models, (i) is equivalent to

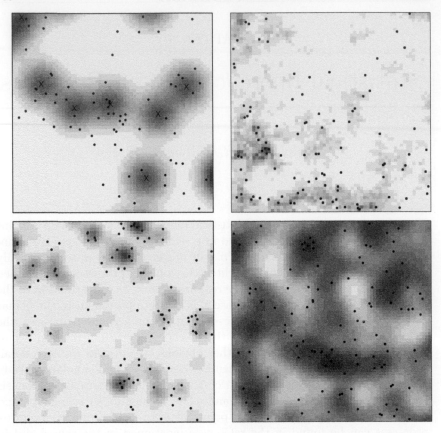

Figure 3.1 *From left to right, top to bottom: realizations of Thomas, LGCP, SNGCP, and "logistic" processes, and associated random intensity surfaces (in gray scale). The crosses in the upper left plot show the cluster centres for the Thomas process. For more details, see Example 4 (Thomas), Example 5 (LGCP), Example 6 (SNGCP), and Example 2 and Example 5 ("logistic").*

a Cox process model (see Section 3.5.1). This is in fact the case for the upper left point pattern in Figure 3.1, where the cluster centres $C = \{c_1, \ldots, c_m\}$ are also shown. The case (ii) is not considered in this contribution, but we refer the interested reader to the literature on Markov point processes, see, for example, Møller (1999) and van Lieshout (2000).

General definitions and descriptions of Poisson and Cox processes are given in Sections 3.2–3.3, while Section 3.4 provides some background on nonparametric analysis. Section 3.5 concerns certain parametric models for Cox processes. Specifically, Section 3.5.1 considers the case where Neyman-Scott processes (Neyman and Scott 1958) are Cox processes, Section 3.5.2

deals with log Gaussian Cox processes (Coles and Jones 1991, Møller et al. 1998), and Section 3.5.3 with shot noise G Cox processes (Brix 1999). The latter class of models includes the Poisson/gamma processes (Wolpert and Ickstadt 1998). As explained in more detail later, the point patterns in Figure 3.1 are realizations of a certain Neyman-Scott process, a log Gaussian Cox process (LGCP), a shot-noise G Cox process (SNGCP), and a certain "logistic process", respectively.

In most applications with an aggregated point pattern modeled by a Cox process X, the underlying environmental process Z is unobserved. Further, only $X \cap W$ is observed, where W is a bounded region contained in the area where the points in X occur. In Section 3.6 we discuss various approaches to estimation in parametric models for Cox processes and focus in particular on minimum contrast estimation. In Section 3.7 we discuss how Z and $X \setminus W$ can be predicted under the various models from Section 3.5. Section 3.8 contains some concluding remarks.

The discussion in Sections 3.5–3.7 will be related to the dataset in Figure 3.2, which shows the positions of 378 weed plants (*Veronica* spp./ speedwell). This point pattern is a subset of a much larger dataset analyzed in Brix and Møller (2001) and Brix and Chadoeuf (2000) where several weed species at different sampling dates were considered. Note that we have rotated the design 90° in Figure 3.2. The 45 frames are of size 30×20 cm^2, and they are organized in 9 groups each containing 5 frames, where the vertical and horizontal distances between two neighbouring groups are 1 m and 1.5 m, respectively. The size of the experimental area is 7.5×5 m^2. The observation window W is given by the union of the 45 frames.

3.2 Poisson Processes

This section surveys some fundamental concepts of point processes and Poisson processes, without going too much into measure theoretical details; for further details, we refer to Daley and Vere-Jones (1988) and Kingman (1993).

By a *point process* X in $\mathbb{R}^d$ we understand a random subset $X \subset \mathbb{R}^d$ which is locally finite, i.e. $X \cap A$ is finite for any bounded region $A \subset \mathbb{R}^d$. By a region we mean a Borel subset of $\mathbb{R}^d$, and measurability of X is equivalent to that

$$N(A) \equiv \text{card}(X \cap A)$$

is a random variable for each bounded region A.

Henceforth X denotes a point process in $\mathbb{R}^d$. Its distribution is determined by the joint distribution of $N(A_1), \ldots, N(A_n)$ for any disjoint regions $A_1, \ldots, A_n$ and any integer $n \geq 1$.

Now, let μ denote an arbitrary locally finite and diffuse measure defined on the regions in $\mathbb{R}^d$, i.e. $\mu(A) < \infty$ for bounded regions A, and $\mu(\{\boldsymbol{x}\}) = 0$

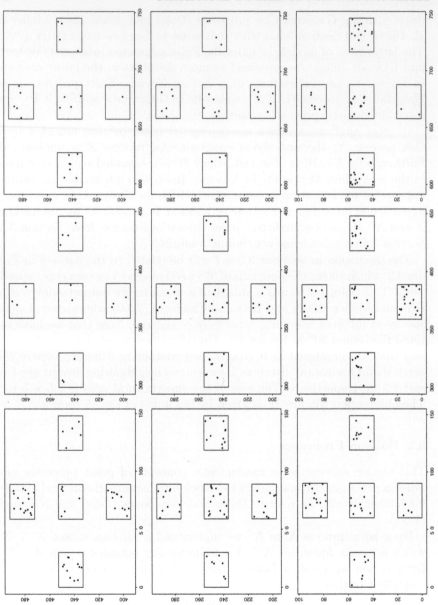

Figure 3.2 *Positions of weed plants when the design is rotated 90°.*

for all singleton sets $\{x\} \subset \mathbb{R}^d$. We say that X is a *Poisson process* with *intensity measure* μ if the following two properties are satisfied:

(a) *independent scattering:* $N(A_1), \ldots, N(A_n)$ are independent for disjoint bounded regions $A_1, \ldots, A_n$ and integers $n \geq 2$;

(b) $N(A)$ is Poisson distributed with mean $\mu(A)$ for bounded regions A.

The properties (a)–(b) are easily seen to be equivalent to (b)–(c), where

(c) for any bounded region A with $\mu(A) > 0$ and any integer $n \geq 1$, conditionally on $N(A) = n$, the n points $x_1, \ldots, x_n$ in $X \cap A$ are independent and each point has distribution $\mu(A \cap \cdot)/\mu(A)$

The conditional distribution in (c) is called a *binomial process*. Using (a)–(c) it is not hard to verify that the Poisson process exists.

A point process is said to be *stationary* if its distribution is invariant under translations in $\mathbb{R}^d$, and *isotropic* if its distribution is invariant under rotations around the origin in $\mathbb{R}^d$. A Poisson process which is stationary is also isotropic, so it is called a *homogeneous* Poisson point process; otherwise it is said to be an *inhomogeneous* Poisson process.

Often μ has a density $\rho : \mathbb{R}^d \to [0, \infty)$ so that $\mu(A) = \int_A \rho(x) dx$ for all regions A. Then ρ is called the *intensity function*. A homogeneous Poisson process has a constant intensity function, and this constant is simply called the *intensity*.

By (b)–(c) it is very easy to simulate a homogeneous Poisson process X on, for example, a rectangular or spherical region B. Denote ρ_{hom} the intensity of X, and imagine we have simulated X on B. Suppose we want to simulate another Poisson process X_{thin} on a bounded region $A \subseteq B$, where X_{thin} has an intensity function ρ_{thin} which is bounded on A by ρ_{hom}. Then we obtain a simulation of $X_{\text{thin}} \cap A$ by including/excluding the points from $X \cap A$ in $X_{\text{thin}} \cap A$ independently of each other, so that a point $x \in X \cap A$ is included in $X_{\text{thin}} \cap A$ with probability $\pi(x) = \rho_{\text{thin}}(x)/\rho_{\text{hom}}$. This procedure is called *independent thinning*; it is extended in Example 2, Sections 3.3–3.4.

Finally, all moments of the counts $N(\cdot)$ for a Poisson process are easily obtained from (a)–(b). For instance,

$$EN(A) = \mu(A) \quad \text{and} \quad \text{cov}(N(A), N(B)) = \mu(A \cap B) \qquad (3.1)$$

for bounded regions A and B.

3.3 Cox Processes

A natural extension of a Poisson process is to let μ be a realization of a random measure M so that the conditional distribution of X given $M = \mu$ follows a Poisson process with intensity measure μ. Then X is said to be a *Cox process driven by* M. This definition can be extended to both multivariate point processes and to space-time processes, see Diggle (1983), Møller et al. (1998), Brix and Møller (2001), and Brix and Diggle (2001).

Example 1 A simple example of a Cox process is a mixed Poisson process with $M(A) = \int_A Z d\boldsymbol{x}$ where Z is a positive random variable, i.e. $X|Z$ follows a homogeneous Poisson process with intensity Z. For example, if Z is gamma distributed, $N(A)$ follows a negative binomial distribution for bounded regions A. Note that $N(A)$ and $N(B)$ are correlated for disjoint bounded regions A and B (except in the trivial case where Z is almost surely constant, i.e. when we have a homogeneous Poisson process).

Example 2 Suppose that X is a Cox process driven by M, and $\Pi = \{\Pi(\boldsymbol{x}) : \boldsymbol{x} \in \mathbb{R}^d\} \subseteq [0,1]$ is a random process which is independent of (X, M). Let X_{thin} denote the point process obtained by random independent thinning of the points in X with retention probabilities Π. More precisely, conditionally on $\Pi = \pi$, X_{thin} is obtained by independent thinning of X where a point in X is retained with probability $\pi(\boldsymbol{x})$. Then X_{thin} is a Cox process driven by $M_{thin}(A) = \int_A \Pi(\boldsymbol{x}) M(d\boldsymbol{x})$. A simple example is shown in the lower right plot in Figure 3.1, where X is a homogeneous Poisson process (i.e. M is proportional to Lebesgue measure), and Π follows a "logistic" process, i.e. $\log(\Pi/(1-\Pi))$ is a Gaussian process (see Example 5 in Section 3.5.2).

Cox processes are like inhomogeneous Poisson processes models for aggregated point patterns. Usually in applications M is unobserved, and so we cannot distinguish a Cox process X from its corresponding Poisson process $X|M$ when only one realization of $X \cap W$ is available (where W denotes the observation window). Which of the two models might be most appropriate, i.e. whether M should be random or "systematic"/deterministic, depends on

- prior knowledge and the scientific questions to be investigated: if, for example, one wants to investigate the dependence of certain covariates associated to M, these may be treated as systematic terms, while unobserved effects may be treated as random terms (for an example, see Benes et al. 2001);

- the particular application: if it seems difficult to model an aggregated point pattern with a parametric class of inhomogeneous Poisson processes (e.g. a class of polynomial intensity functions), Cox process models such as those in Section 3.5 may allow more flexibility and/or a more parsimonious parametrization;

- another application is nonparametric Bayesian modelling: nonparametric Bayesian smoothing for the intensity surface of an inhomogeneous Poisson process is treated in Heikkinen and Arjas (1998) (whose approach is very similar to the one in Chapter 6); see also Remark 4 in Section 3.8.

Distributional properties of a Cox process X driven by M follow immediately by conditioning on M and exploiting the properties of the Poisson process $X|M$. For instance, by (3.1),

$$EN(A) = EM(A)$$

and

$$\mathrm{cov}(N(A), N(B)) = \mathrm{cov}(M(A), M(B)) + EM(A \cap B)$$

for bounded regions A and B. Hence, $\mathrm{var}(N(A)) = \mathrm{var}(M(A)) + EM(A) \geq EN(A)$ with equality only when $M(A)$ is almost surely constant as in the Poisson case. In other words, a Cox process exhibits over dispersion when compared to a Poisson process.

In many specific models for Cox processes, including those considered in Examples 1–2 and in Section 3.5, M is specified by a nonnegative spatial process $Z = \{Z(\boldsymbol{x}) : \boldsymbol{x} \in \mathbb{R}^d\}$ so that

$$M(A) = \int_A Z(\boldsymbol{x}) \mathrm{d}\boldsymbol{x}. \tag{3.2}$$

Then we say that X is *driven by the random intensity surface Z*. In the sequel we restrict attention to Cox processes driven by a random intensity surface.

Simulation of X is easy in principle: if we have a simulation $z_A = \{z(\boldsymbol{x}) : \boldsymbol{x} \in A\}$ of Z restricted to a bounded region A, where z_A is bounded by a constant, then the simulation method at the end of Section 3.2 can be used to obtain a realization of $X \cap A|Z_A = z_A$. Note that by the independent scattering property (a) in Section 3.2, the conditional distribution $X \cap A|Z$ depends only on Z_A.

3.4 Summary Statistics

The first and second order moments of the counts $N(\cdot)$ for a Cox process can be expressed in terms of two functions:

$$\rho(\boldsymbol{x}) = E[Z(\boldsymbol{x})] \quad \text{and} \quad g(\boldsymbol{x}_1, \boldsymbol{x}_2) = E[Z(\boldsymbol{x}_1)Z(\boldsymbol{x}_2)]/[\rho(\boldsymbol{x}_1)\rho(\boldsymbol{x}_2)]$$

which are called the *intensity function* and the *pair correlation function*, respectively ("pair correlation function" is standard terminology, though it is somewhat misleading). The intensity and pair correlation functions can be defined for general point process models and not just Cox processes, see, for example, Stoyan et al. (1995). Intuitively, $E[Z(\boldsymbol{x}_1)Z(\boldsymbol{x}_2)]\mathrm{d}\boldsymbol{x}_1\mathrm{d}\boldsymbol{x}_2$ is the probability for having a point from X in each of the infinitesimally small volumes $\mathrm{d}\boldsymbol{x}_1$ and $\mathrm{d}\boldsymbol{x}_2$. It is, however, more informative to work with the pair correlation function which is normalized with the "marginal probabilities" $\rho(\boldsymbol{x}_1)\mathrm{d}\boldsymbol{x}_1$ and $\rho(\boldsymbol{x}_2)\mathrm{d}\boldsymbol{x}_2$ for observing a point in $\mathrm{d}\boldsymbol{x}_1$ and $\mathrm{d}\boldsymbol{x}_2$, respectively. Another good reason is that g is invariant under random independent thinning of a point process; this is exemplified in Example 3 below for the particular case of a Cox process. Finally, $g = 1$ in the Poisson case.

Example 3 Consider again Example 2 with M given by (3.2), and let ρ and g denote the intensity and pair correlation functions of X. Then X_{thin} is a Cox process with random intensity surface $Z_{thin}(x) = \Pi(x)Z(x)$, intensity function $\rho_{thin}(x) = E\Pi(x)\rho(x)$, and pair correlation function $g_{thin} = g$.

The intensity and pair correlation functions can be estimated by non-parametric methods under rather general conditions, see Diggle (1985b), Stoyan & Stoyan (1994, 2000), Baddeley et al. (2000), Ohser and Mücklich (2000), and Møller and Waagepetersen (2001). When the pair correlation function exists (as it does for the Cox processes we consider), all we need to assume is that g is invariant under translations, i.e. $g(x_1, x_2) = g_0(x_1 - x_2)$. Then also a so-called K-*function* can be defined by

$$K(r) = \int_{\|x\| \le r} g_0(x)\mathrm{d}x, \quad r \ge 0, \tag{3.3}$$

and it is easier to estimate K than g by nonparametric methods (Baddeley et al. 2000). For a stationary point process X, Ripley's K-function (Ripley 1976) agrees with K in (3.3), and $\rho K(r)$ can be interpreted as the expected number of further points within distance r from a typical point in X. Note that K and g_0 are in one-to-one correspondence if $g_0(x)$ depends only on the distance $\|x\|$. Moreover, one often makes the transformation

$$L(r) = [K(r)/(\pi^{d/2}/\Gamma(1 + d/2))]^{1/d}$$

as $L(r) = r$ is the identity in the Poisson case.

Nonparametric estimators of ρ, g, K, and L are *summary statistics* of the first and second order properties of a spatial point process. These may be supplied with other summary statistics, including nonparametric estimators of the so-called F, G, and J functions which are based on interpoint distances for a stationary point process X. Briefly, for any $r > 0$, $1 - F(r)$ is the probability that X has no points within distance r from an arbitrary fixed point in $\mathbb{R}^d$, $1 - G(r)$ is the conditional probability that X has no further points within distance r from a typical point in X, and $J(r) = (1 - G(r))/(1 - F(r))$ (defined for $F(r) < 1$; see van Lieshout & Baddeley 1996). For a stationary Cox process,

$$1 - F(r) = E \exp\left[\left(-\int_{\|x\| \le r} Z(x)\mathrm{d}x\right)\right],$$

and

$$1 - G(r) = E\left[\exp\left(-\int_{\|x\| \le r} Z(x)\mathrm{d}x\right)Z(0)\right]\Big/\rho.$$

In the special case of a homogeneous Poisson process,

$$1 - F(r) = 1 - G(r) = \exp\left(-\rho r^d \pi^{d/2}/\Gamma(1 + d/2)\right).$$

Further results are given in Section 3.5.1.

A plot of a summary statistic is often supplied with *envelopes* obtained by simulation of a point process under some specified model: Consider, for example, the L-function. Let $\hat{L}_0$ be a nonparametric estimator of L obtained from X observed within some window W, and let $\hat{L}_1, \ldots, \hat{L}_n$ be estimators obtained in the same way as $\hat{L}_0$ but from i.i.d. simulations $X_1, \ldots, X_n$ under the specified model for X. Then, for each distance r, we have that

$$\min_{1 \leq i \leq n} \hat{L}_i(r) \leq \hat{L}_0 \leq \max_{1 \leq i \leq n} \hat{L}_i(r) \qquad (3.4)$$

with probability $(n-1)/(n+1)$ if X follows the specified model. We refer to the bounds in (3.4) as lower and upper envelopes. In our examples we choose $n = 39$ so that (3.4) specifies a 2.5% lower envelope and a 97.5% upper envelope.

The estimated L, g, and F functions for the weed data in Figure 3.2 are shown in Figures 3.3 and 3.4. The figures also show the averages of these summary statistics and envelopes obtained from 39 simulations under a homogeneous Poisson process with expected number of points in W equal to the observed number of points. The plots for L and g clearly indicate aggregation, so the homogeneous Poisson process provides a poor fit. We do not consider G due to problems with handling of edge effects in the special experimental design for the weed plants (see the discussion in Brix & Møller, 2001). The averages are close to the theoretical curves except for g where $\hat{g}(r)$ is biased upwards for small $r < 2.5$ cm (see the discussion p. 286 in Stoyan & Stoyan, 1994). The envelopes for g are rather wide for 25 cm $< r <$ 55 cm where few interpoint distances are observed. Similarly, the envelopes for L are wide for $r > 25$ cm.

3.5 Parametric Models of Cox Processes

3.5.1 Neyman-Scott Processes as Cox Processes

A *Neyman-Scott process* (Neyman and Scott 1958) is a particular case of a *Poisson cluster process*. In this section we consider the case when Neyman-Scott processes become Cox processes.

Let C be a homogeneous Poisson process of cluster centres with intensity $\kappa > 0$. Assume that conditionally on $C = \{c_1, c_2, \ldots\}$, the clusters $X_1, X_2, \ldots$ are independent Poisson processes, where the intensity measure of X_i, $i \geq 1$, is given by

$$\mu_i(A) = \int_A \alpha f(\boldsymbol{x} - \boldsymbol{c}_i) \mathrm{d}\boldsymbol{x} \qquad (3.5)$$

where $\alpha > 0$ is a parameter and f is a density function for a continuous random variable in $\mathbb{R}^d$. Then $X = \cup_i X_i$ is a Neyman-Scott process. It is

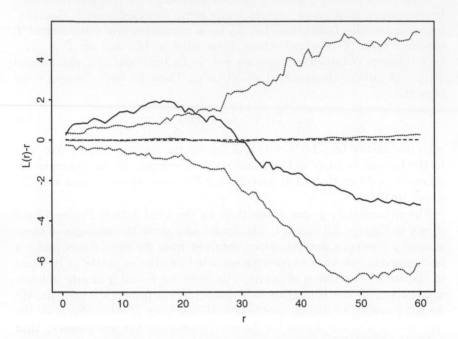

Figure 3.3 *Estimated $L(r) - r$ for positions of weed plants (solid line), envelopes, and average calculated from 39 simulations under the fitted homogeneous Poisson process (short-dashed line), and theoretical value of $L(r) - r$ (long-dashed line).*

also a Cox process with random intensity surface

$$Z(\boldsymbol{x}) = \sum_{\boldsymbol{c} \in C} \alpha f(\boldsymbol{x} - \boldsymbol{c}). \qquad (3.6)$$

The process (3.6) is stationary. It is also isotropic if $f(\boldsymbol{x})$ only depends on $\|\boldsymbol{x}\|$. The intensity and the pair correlation function are given by

$$\rho = \alpha\kappa, \quad g_0(\boldsymbol{x}) = 1 + h(\boldsymbol{x})/\kappa, \qquad (3.7)$$

where

$$h(\boldsymbol{x}) = \int f(\boldsymbol{x}_0) f(\boldsymbol{x} + \boldsymbol{x}_0) \mathrm{d}\boldsymbol{x}_0$$

is the density for the difference between two independent points which each have density f. Furthermore,

$$J(r) = \int f(\boldsymbol{x}_1) \exp\left(-\alpha \int_{\|\boldsymbol{x}_2\| \le r} f(\boldsymbol{x}_1 + \boldsymbol{x}_2) \mathrm{d}\boldsymbol{x}_2\right) \mathrm{d}\boldsymbol{x}_1,$$

see Bartlett (1975) and van Lieshout & Baddeley (1996). Hence, $F \le G$ and J is nonincreasing with range $[\exp(-\alpha), 1]$.

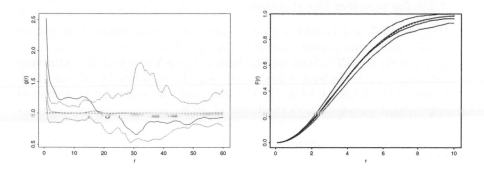

Figure 3.4 *Left and right: as Figure 3.3 but for g and F, respectively.*

Example 4 Closed form expressions for g_0 are known for a few Neyman-Scott models. A Thomas process (Thomas 1949) has

$$f(\boldsymbol{x}) = \left(2\pi\omega^2\right)^{-d/2} \exp\left(-\|\boldsymbol{x}\|^2/(2\omega^2)\right), \tag{3.8}$$

the density for $N_d(0, \omega^2 I_d)$, i.e. for d independent normally distributed variables with mean 0 and variance $\omega^2 > 0$. This process is isotropic with

$$g_0(\boldsymbol{x}) = 1 + \left(4\pi\omega^2\right)^{-d/2} \exp\left(-\|\boldsymbol{x}\|^2/(4\omega^2)\right)/\kappa. \tag{3.9}$$

The K-function can be expressed in terms of the cumulative distribution function for a χ^2-distribution with d degrees of freedom, and for $d = 2$ we simply obtain

$$K(r) = \pi r^2 + [1 - \exp(-r^2/(4\omega^2))]/\kappa. \tag{3.10}$$

The upper left plot in Figure 3.1 shows a simulation of a Thomas process with $\kappa = 10$, $\alpha = 10$, and $\omega^2 = 0.1$.

Another mathematically tractable model is a Matérn cluster process (Matérn 1960, Matérn 1986), where f is the uniform density on a d-dimensional ball with centre at the origin; see Santaló (1976) and Stoyan et al. (1995).

We can obviously modify the definition of a Neyman-Scott process and the results above in many ways, for example, by replacing α and $f(\boldsymbol{x} - \boldsymbol{c}_i)$ in (3.5) by $\alpha(\boldsymbol{c}_i)$ and $f_{\boldsymbol{c}_i}(\boldsymbol{x})$ (using an obvious notation), or by replacing C in (3.6) by an arbitrary Poisson process — we consider such extensions in Section 3.5.3. In either case we obtain a Cox process.

Simulation of a Neyman-Scott process follows in principle straightforwardly from either (3.6) and its definition as a Cox process or from its construction as a Poisson cluster process. However, boundary effects play a role: if we wish to simulate X within a bounded region W, we first simulate C within an extended region $B_{\mathrm{ext}} \supset W$ so that points in X associated to cluster centres in $C \setminus B_{\mathrm{ext}}$ fall in W with a negligible probability.

3.5.2 Log Gaussian Cox Processes

If $Y = \log Z$ is a Gaussian process, i.e. if any finite linear combination $\sum a_i Y(x_i)$ follows a normal distribution, then X is said to be a *log Gaussian Cox process (LGCP)*. Such models have independently been introduced in astronomy by Coles and Jones (1991) and in statistics by Møller et al. (1998). The definition of an LGCP can easily be extended in a natural way to multivariate LGCPs (Møller et al. 1998) and to multivariate spatio-temporal LGCPs (Brix and Møller 2001, Brix and Diggle 2001).

It is necessary to impose weak conditions on the mean and covariance functions

$$m(x) = E[Y(x)] \quad \text{and} \quad c(x_1, x_2) = \text{cov}(Y(x_1), Y(x_2))$$

in order to get a well-defined and finite integral $\int_B Z(x)dx$ for bounded regions B. For example, we may require that $x \to Y(x)$ is almost surely continuous. Weak conditions ensuring this, and which are satisfied for the models of m and c used in practice, are given in Theorem 3.4.1 in Adler (1981) (or see Møller et al. 1998).

Example 5 The covariance function belongs to the power exponential family if

$$c(x_1, x_2) = \sigma^2 \exp\left(-\|(x_1 - x_2)/\alpha\|^\delta\right), \quad 0 \le \delta \le 2, \tag{3.11}$$

where $\alpha > 0$ is a scale parameter for the correlation and $\sigma^2 > 0$ is the variance. The special case $\delta = 1$ is an exponential covariance function, and $\delta = 2$ a Gaussian covariance function. If m is continuous, then $x \to Y(x)$ is almost surely continuous. For the LGCP in the upper right plot in Figure 3.1, $(\delta, \alpha) = (1, 0.14)$. The lower right plot in Figure 3.1 is for the logistic process in Example 2 with $(\delta, \alpha) = (2, 0.10)$. In both plots the Gaussian process has mean zero and variance one.

Due to the richness of possible mean and covariance functions, LGCPs are flexible models for aggregation as demonstrated in Møller et al. (1998), where examples of covariance functions together with simulated realizations of LGCPs and their underlying Gaussian processes are shown. In the specific examples of applications considered in this chapter, we let m be constant and c an exponential covariance function. Brix and Møller (2001) and Møller and Waagepetersen (2001) consider situations where m is a linear or polynomial function, and Benes et al. (2001) consider a case where m depends on covariates.

The intensity and pair correlation functions of an LGCP are given by

$$\rho(x) = \exp(m(x) + c(x, x)/2) \quad \text{and} \quad g(x_1, x_2) = \exp(c(x_1, x_2)). \tag{3.12}$$

Hence there is a one-to-one correspondence between (m, c) and (ρ, g), and the distribution of an LGCP is determined by (ρ, g). This makes parametric models easy to interpret. Moreover, stationarity respective isotropy of an LGCP is equivalent to stationarity respective isotropy of (m, c) or equivalently of (ρ, g).

The simple relationship (3.12) indicates that LGCPs are more tractable for mathematical analysis than Neyman-Scott processes; further results are given in Møller et al. (1998).

We now turn to simulation of an LGCP. In contrast to the case of Neyman-Scott processes, we do not have problems with boundary effects since the distribution of an LGCP restricted to a bounded region B depends only on the distribution of $Y_B = \{Y(\boldsymbol{x}) : \boldsymbol{x} \in B\}$. As Y_B does not in general have a finite representation in a computer, we approximate Y_B by a random step function with constant value $Y(\boldsymbol{c}_i)$ within disjoint cells C_i. Here $B = \cup_{i \in I} C_i$ is a subdivision with finite index set I, and $\boldsymbol{c}_i \in C_i$ is a "centre" point of C_i. So we actually consider how to simulate the Gaussian vector $\tilde{Y} = (\tilde{Y}_i)_{i \in I}$ where $\tilde{Y}_i = Y(\boldsymbol{c}_i)$.

As discussed in Møller et al. (1998), there is an efficient way of simulating $\tilde{Y}$ when $c(\xi, \eta) = c(\xi - \eta)$ is invariant under translations, $d = 2$, and B is rectangular: Let $I \subset B$ denote a rectangular grid which is embedded in a rectangular grid I_{ext}, which is wrapped on a torus. Then a block circulant matrix $K = \{K_{ij}\}_{i,j \in I_{\text{ext}}}$ can be constructed so that $\{K_{ij}\}_{(i,j) \in I}$ is the covariance matrix of $\tilde{Y}$. Since K is block circulant, it can easily be diagonalized by means of the two-dimensional discrete Fourier transform with associated matrix F_2 (see Wood & Chan 1994 and Section 6.1 in Møller et al. 1998). Suppose that K has non-negative eigenvalues. Then we can extend $\tilde{Y} = (\tilde{Y}_{(i,j)})_{(i,j) \in I}$ to a larger Gaussian field $\tilde{Y}_{\text{ext}} = (\tilde{Y}_{(i,j)})_{(i,j) \in I_{\text{ext}}}$ with covariance matrix K: set

$$\tilde{Y}_{\text{ext}} = \Gamma Q + \mu_{\text{ext}} \tag{3.13}$$

where Γ follows a standard multivariate normal distribution, $Q = \bar{F}_2 \Lambda^{1/2} F_2$, Λ is the diagonal matrix of eigenvalues for K, and the restriction of μ_{ext} to I agrees with the mean of $\tilde{Y}$. Using the two-dimensional fast Fourier transform a fast simulation algorithm for $\tilde{Y}_{\text{ext}}$ and hence $\tilde{Y}$ is obtained. We use this method for the simulations in Figure 3.1 and in connection with the MCMC algorithm considered in Section 3.7.2.

Another possibility is to use the Choleski decomposition of the covariance matrix of $\tilde{Y}$, provided this covariance matrix is positive definite. This may be advantageous if the covariance function is not translation invariant or B is far from being rectangular. On the other hand, the Choleski decomposition is only practically applicable if the dimension of $\tilde{Y}$ is moderate.

3.5.3 Shot Noise G Cox Processes

Brix (1999) introduces *shot noise G Cox processes (SNGCP)* by smoothing a so-called G measure which is driving a so-called G Cox process, see Remark 2 in Section 3.8. We instead define directly a SNGCP as a Cox process X with

$$Z(\boldsymbol{x}) = \sum_j \gamma_j k(\boldsymbol{x}, \boldsymbol{c}_j) \qquad (3.14)$$

where the notation means the following. For each $\boldsymbol{c} \in \mathbb{R}^p$, $k(\cdot, \boldsymbol{c})$ is a density function for a continuous random variable. Further, $\{(\boldsymbol{c}_j, \gamma_j)\}$ is a Poisson process defined on $\mathbb{R}^p \times [0, \infty)$ by the intensity measure

$$\nu(A \times B) = (\kappa(A)/\Gamma(1-\alpha)) \times \int_B \gamma^{-\alpha-1} \exp(-\tau\gamma) \mathrm{d}\gamma, \quad A \subseteq \mathbb{R}^p,\ B \subseteq [0, \infty).$$
$$(3.15)$$

Here $\alpha < 1$ and $\tau \geq 0$ are parameters with $\tau > 0$ if $\alpha \leq 0$, and κ is a locally finite measure.

This definition can obviously be modified in many ways. For instance, if $\alpha = 0$ and we redefine ν by

$$\nu(A \times B) = (1/\Gamma(1-\alpha)) \int_A \int_B \gamma^{-\alpha-1} \exp(-\tau(\boldsymbol{c})\gamma) \kappa(\mathrm{d}\boldsymbol{c}) \mathrm{d}\gamma,$$

where τ is now a positive measurable function, we obtain a Poisson/gamma process as studied in Wolpert and Ickstadt (1998). Extensions to multivariate SNGCPs are also obvious (Brix 1999). In the sequel X denotes an SNGCP with Z and ν given by (3.14) and (3.15).

Assume that $\kappa(\cdot) = \kappa_0 |\cdot|$ is proportional to Lebesgue measure. Then $\{\boldsymbol{c}_j\}$ is stationary. Assume also that the kernels $k(\boldsymbol{x}, \boldsymbol{c}) = k_0(\boldsymbol{x} - \boldsymbol{c})$, $\boldsymbol{x}, \boldsymbol{c} \in \mathbb{R}^p$, are given by a common kernel k_0. Then X is stationary, and the intensity and pair correlation functions exist for $\tau \neq 0$ and are given by

$$\rho = \kappa_0 \tau^{\alpha-1}, \quad g_0(\boldsymbol{x}) = 1 + \frac{1-\alpha}{\kappa_0 \tau^\alpha} \int k_0(\boldsymbol{x}_0) k_0(\boldsymbol{x} + \boldsymbol{x}_0) \mathrm{d}\boldsymbol{x}_0. \qquad (3.16)$$

These are of the same form as for a Neyman-Scott process, cf. (3.7), so we cannot distinguish between the two types of models by considering a nonparametric estimate of (ρ, g_0) only. If furthermore $k_0(\boldsymbol{x})$ depends only on the distance $\|\boldsymbol{x}\|$, then X is isotropic.

Example 6 If k_0 is given by the normal density f in (3.8) with variance ω^2, X is isotropic, and g and K are of the same form as for a Thomas process (replacing κ in (3.9) and (3.10) by $\tau^\alpha \kappa_0/(1-\alpha)$). In the lower left plot in Figure 3.1, $\alpha = 0$, $\kappa_0 = 50$, $\tau = 0.5$ and $\omega^2 = 0.001$.

The marginal distributions of the independent processes $\{\boldsymbol{c}_j\}$ and $\{\gamma_j\}$ depend much on α as described below.

The case $\alpha < 0$: Then $\{\boldsymbol{c}_j\}$ is a Poisson process with intensity measure $(\tau^\alpha/|\alpha|)\kappa$, and the γ_j are mutually independent and follow a common

Gamma distribution $\Gamma(|\alpha|, \tau)$. Hence, X is a kind of modified Neyman-Scott process. Particularly, X can be simulated within a bounded region in much the same way as we would simulate a Neyman-Scott process, cf. Section 3.5.1.

The case $0 \leq \alpha < 1$: The situation is now less simple as $\{c_j\}$ is not locally finite. For $\alpha = 0$, we have a Poisson/gamma model (Daley and Vere-Jones 1988, Wolpert and Ickstadt 1998). For simplicity, suppose that κ is concentrated on a region D with $0 < \kappa(D) < \infty$, and write $\{(c_j, \gamma_j)\} = \{(c_1, \gamma_1), (c_2, \gamma_2), \ldots\}$ so that $\gamma_1 > \gamma_2 > \ldots > 0$. Define $q(t) = \nu(D \times [t, \infty))$ for $t > 0$. Then q is a strictly decreasing function which maps $(0, \infty)$ onto $(0, \infty)$, and $q(\gamma_1) < q(\gamma_2) < \ldots$ are the points of a homogeneous Poisson process of intensity 1 and restricted to $(0, \infty)$. Furthermore, $c_1, c_2, \ldots$ are independent, and each c_j follows the probability measure $\kappa(\cdot)/\kappa(D)$. For simulation and inference, one approximates Z by

$$Z(\boldsymbol{x}) \approx Z^J(\boldsymbol{x}) = \sum_{j=1}^{J} \gamma_j k(\boldsymbol{x}, \boldsymbol{c}_j) \qquad (3.17)$$

where $J < \infty$ is a "cut off". Brix (1999) and Wolpert and Ickstadt (1998) discuss how q^{-1} and the tail sum $\sum_{j>J} \gamma_j$ can be evaluated.

3.6 Estimation for Parametric Models of Cox Processes

For specificity we discuss first estimation in the context of a Thomas process X with unknown parameters $\kappa > 0$, $\alpha > 0$, and $\sigma > 0$. We next turn to LGCPs and SNGCPs in Examples 8 and 9.

In most applications the process of cluster centres C is treated as missing data, and we have only observed $X \cap W = \{\boldsymbol{x}_1, \ldots, \boldsymbol{x}_n\}$, where W is a bounded observation window. Let $B_{\text{ext}} \supset W$ be specified as in the end of Section 3.5.1, and redefine the random intensity surface (3.6) so that the sum over cluster centres C is replaced by a sum over $C \cap B_{\text{ext}}$.

The *likelihood function* for $\theta = (\kappa, \alpha, \omega)$ is

$$L(\theta) = E_\kappa \left[\exp \left(- \int_W Z(\boldsymbol{x}; C \cap B_{\text{ext}}, \alpha, \omega) \mathrm{d}\boldsymbol{x} \right) \prod_{j=1}^{n} Z(\boldsymbol{x}_j; C \cap B_{\text{ext}}, \alpha, \omega) \right]$$
$$(3.18)$$

where the mean is with respect to the Poisson process $C \cap B_{\text{ext}}$, and

$$Z(\boldsymbol{x}; C \cap B_{\text{ext}}, \alpha, \omega) = (\alpha/\omega^2) \sum_{\boldsymbol{c} \in C \cap B_{\text{ext}}} \varphi((\boldsymbol{x} - \boldsymbol{c})/\omega)$$

where φ denotes the density for the d-dimensional standard normal distribution. If W is rectangular (or a disjoint union of rectangular sets as in Figure 3.2), the integral in (3.18) can be expressed in terms of the cumulative distribution function for $N(0, 1)$. The likelihood (3.18) can further be estimated by *Markov chain Monte Carlo* (MCMC) methods: For a given value $\theta_0 = (\kappa_0, \alpha_0, \omega_0)$ of θ, suppose $C^1 = \{c_1^1, \ldots, c_{m_1}^1\}, \ldots, C^k = \{c_1^k, \ldots, c_{m_k}^k\}$ is a Markov chain sample from the conditional distribution $C \cap B_{\mathrm{ext}} | X \cap W = \{x_1, \ldots, x_n\}$ (Section 3.7.1 describes how such a sample can be generated). The Monte Carlo approximation of $L(\theta)/L(\theta_0)$ is

$$\frac{1}{k} \sum_{i=1}^{k} \frac{\kappa^{m_i} \exp\left(-\int_W (\kappa + Z(x; C^i, \alpha, \omega)) \mathrm{d}x\right) \prod_{j=1}^{n} Z(x_j; C^i, \alpha, \omega)}{\kappa_0^{m_i} \exp\left(-\int_W (\kappa_0 + Z(x; C^i, \alpha_0, \omega_0)) \mathrm{d}x\right) \prod_{j=1}^{n} Z(x_j; C^i, \alpha_0, \omega_0)}.$$

Note that the approximation is based on importance sampling, so it only works for θ sufficiently close to θ_0. The generation and storing of $C^1, \ldots, C^k$ is further computationally rather demanding; see also Remark 1 in Section 3.8 and Chapter 4.

A computationally easier alternative for parameter estimation is *minimum contrast estimation* (Diggle 1983, Møller et al. 1998): The closed form expressions (3.9) and (3.10) may be compared with the nonparametric summary statistics $\hat{g}$ and $\hat{K}$ obtained assuming only stationarity and isotropy of X (Section 3.4). If for example $d = 2$ and the K-function is used, a minimum contrast estimate $(\hat{\kappa}, \hat{\omega})$ is chosen to minimize

$$\int_{a_1}^{a_2} \left\{ (\hat{K}(r) - \pi r^2) - \left([1 - \exp(-r^2/(4\omega^2))]/\kappa\right) \right\}^2 \mathrm{d}r \qquad (3.19)$$

where $0 \le a_1 < a_2$ are user specified parameters (see Example 7 below). Setting

$$A(\omega^2) = \int_{a_1}^{a_2} [1 - \exp(-r^2/(4\omega^2))]^2 \mathrm{d}r$$

and

$$B(\omega^2) = \int_{a_1}^{a_2} [(\hat{K}(r) - \pi r^2)(1 - \exp(-r^2/(4\omega^2)))] \mathrm{d}r,$$

then

$$\hat{\omega}^2 = \arg\max\left[B(\omega^2)^2/A(\omega^2)\right], \quad \hat{\kappa} = \left[A(\hat{\omega}^2)/B(\hat{\omega}^2)\right].$$

Inserting this into the equation $\rho = \alpha\kappa$ and using the natural estimate $\hat{\rho} = n/|W|$ where $|W|$ denotes the area of W, we obtain finally the estimate

$$\hat{\alpha} = \hat{\rho}/\hat{\kappa}.$$

Diggle (1983) suggests the alternative contrast

$$\int_{a_1}^{a_2} \left\{ (\hat{K}(r))^b - (\pi r^2 + [1 - \exp(-r^2/(4\omega^2))]/\kappa)^b \right\}^2 \mathrm{d}r$$

where $b > 0$ is a third user-specified parameter recommended to be between 0.25 and 0.5. This approach requires numerical minimization with respect to κ and ω jointly. Brix (1999) reports that minimum contrast estimation for SNGCPs is numerically more stable if g is used instead of K. Minimum contrast estimation for LGCPs and space-time LGCPs is considered in

Møller et al. (1998), Brix and Møller (2001), and Brix and Diggle (2001); see also Example 8 below.

Example 7 (Thomas) For the weed data and certain choices of a_1 and a_2, the contrast (3.19) did not have a well-defined minimum. Numerically more stable results were obtained by using the g-function given by (3.9), i.e. by considering the contrast obtained by replacing $\hat{K}(r) - \pi r^2$ and $[1 - \exp(-r^2/(4\omega^2))]/\kappa$ in (3.19) by $\hat{g}(r) - 1$ and $\exp(-r^2/(4\omega^2)/(4\pi\omega^2\kappa)$, respectively. The middle plot in Figure 3.4 suggests to use $a_1 = 2.5$ cm and $a_2 = 25$ cm due to the bias of $\hat{g}(r)$ for $r < 2.5$ cm and the large variability of $\hat{g}(r)$ for $r > 25$ cm. With these values of a_1 and a_2 the estimates $\hat{\kappa}=0.005$, $\hat{\omega}^2 = 51.0$, and $\hat{\alpha} = 3.05$ are obtained.

Example 8 (LGCP) Turning next to a stationary LGCP, the mean function m is equal to a real constant ξ, say. Let $c = c_{\alpha,\sigma}$ be the exponential covariance function with positive parameters (α, σ) as in (3.11) with $\delta = 1$. For similar reasons as for the Thomas process, likelihood estimation of $\theta = (\xi, \alpha, \sigma^2)$ is computationally demanding, and minimum contrast estimation is a simpler method. Because of the simple relationship (3.12), the minimum contrast estimate $(\hat{\alpha}, \hat{\sigma}^2)$ is chosen to minimize

$$\int_{a_1}^{a_2} \left\{ \hat{g}(r)^b - \left(\sigma^2 \exp(-r/\alpha) \right)^b \right\}^2 dr$$

where $b > 0$. As for the Thomas process we can easily find $(\hat{\alpha}, \hat{\sigma}^2)$ and combine this with (3.12) and the estimate for the intensity to obtain the equation $n/|W| = \exp(\hat{\xi} + \hat{\sigma}^2/2)$ for the estimate of ξ. Using the same values of a_1 and a_2 as in Example 5 and letting $b = 1$, we obtain the minimum contrast estimates $\hat{\xi} = -4.5$, $\hat{\sigma}^2 = 0.45$, and $\hat{\alpha} = 9.5$.

Example 9 (SNGCP) Consider the SNGCP from Example 6. Using minimum contrast estimation as for the Thomas process we obtain $\hat{\kappa}_0=0.005$, $\hat{\omega}^2 = 51.0$, and $\hat{\tau} = 0.33$.

A comparison of the nonparametric estimates $\hat{g}$ and $\hat{F}$ with the envelopes calculated under the various fitted models in Examples 7–9 (see Figure 3.5) does not reveal obvious inconsistencies between the data and any of the fitted models, except perhaps for the SNGCP where $\hat{F}(r)$ coincides with the upper envelope for a range of distances r. The bias of $\hat{g}$ near zero makes it difficult to make inference about the behaviour of the pair correlation near zero. For an LGCP it is for example difficult to infer whether an exponential or a "Gaussian" covariance function should be used for the Gaussian field Y. See also Remark 1, Section 3.8.

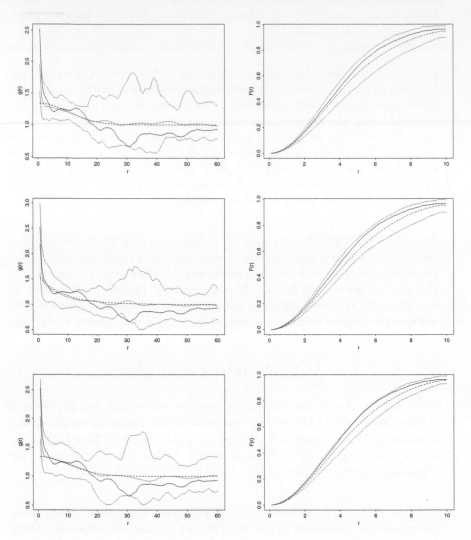

Figure 3.5 *Upper left: $\hat{g}(r)$ for the weed data (solid line), envelopes and average calculated from 39 simulations under the fitted Thomas process (short-dashed line), and theoretical value of $g(r)$ for fitted Thomas process (long-dashed line). Upper right: estimated $F(r)$ (solid line), and envelopes and average calculated from 39 simulations under the fitted Thomas process (short-dashed line). Middle plots: as upper plots but for LGCP (Example 8). Lower plots: as upper plots but for SNGCP (Example 9).*

3.7 Prediction

As in Section 3.6 suppose that a realization $x = \{\boldsymbol{x}_1, \ldots, \boldsymbol{x}_n\}$ of $X \cap W$ is observed, where W is a bounded observation window. Given a bounded region

B, one may be interested in predicting $Z_B = \{Z(x)\}_{x \in B}$ or $X \cap (B \setminus W)$ or, more generally, in computing the conditional distributions of Z_B and $X \cap (B \setminus W)$ given $X \cap W = x$. In the sequel we mainly focus on the conditional distribution of Z_B, since $X \cap (B \setminus W)$ is conditionally independent of $X \cap B$ given Z_B, and $X \cap (B \setminus W)$ is simply an inhomogeneous Poisson process given Z_B. The conditional distribution of Z_B is in general not analytically tractable, so MCMC methods become useful for computing characteristics of the conditional distribution. In the following we discuss MCMC algorithms for the model classes considered in Section 3.5. For background knowledge on MCMC algorithms (particularly Metropolis-Hastings algorithms), see, for example, Gilks et al. (1996) and Robert and Casella (1999).

3.7.1 Conditional Simulation for Neyman-Scott Processes

For a Neyman-Scott process (Section 3.5.1), Z_B is determined by the process of cluster centres. We approximate this by the process of cluster centres falling in a sufficiently large region B_{ext} which contains B, and ignore cluster centres outside B_{ext}. The conditional distribution of the cluster centres on B_{ext} then has a density with respect to the homogeneous Poisson process of intensity 1 and restricted to B_{ext}. The density is given by

$$p(c|x) \propto \kappa^{\text{card}(c)} \exp\left(-\int_W Z(x)dx \right) \prod_{c \in c} Z(c)$$

for finite point configurations $c \subset B_{\text{ext}}$, where $Z(x)$ is given by (3.6) but with C replaced by c. The conditional distribution can be simulated using the MCMC algorithm studied in Geyer and Møller (1994). Prediction and Bayesian inference for Neyman-Scott processes viewed as cluster processes has been considered by, for example, Lawson (1993a), Baddeley and van Lieshout (1993), and van Lieshout & Baddeley (1995); see also Chapters 4 and 5.

3.7.2 Conditional Simulation for LGCPs

Consider now an LGCP with $Y_B = \log Z_B$ where we use a notation as in Section 3.5.2. Approximate simulations of $Y_B|X_W = x$ can be obtained from simulations of $\tilde{Y}|X_W = x$, which in turn can be obtained from simulations of $\Gamma|X_W = x$ using the transformation (3.13). Omitting an additive constant depending on x only, the log conditional density of Γ given x is

$$-\|\gamma\|^2/2 + \sum_{i \in I}(\tilde{y}_i n_i - A_i \exp(\tilde{y}_i)) \tag{3.20}$$

where, in accordance with (3.13), $(\tilde{y}_i)_{i \in I_{\text{ext}}} = \gamma Q + \mu_{\text{ext}}$, $n_i = \text{card}(x \cap C_i)$, and $A_i = |C_i|$ if $C_i \subset W$ and $A_i = 0$ otherwise. The gradient of (3.20) becomes

$$\nabla(\gamma) = -\gamma + \left(n_i - A_i \exp(\tilde{y}_i)\right)_{i \in I_{\text{ext}}} Q^{\mathsf{T}},$$

and differentiating once more, the conditional density of Γ given x is seen to be strictly log-concave.

For simulation from $\Gamma | X_W = x$, Møller et al. (1998) use a *Langevin-Hastings algorithm* or *Metropolis-adjusted Langevin algorithm* as introduced in the statistical community by Besag (1994) (see also Roberts & Tweedie 1996) and earlier in the physics literature by Rossky et al. (1978). This is a Metropolis-Hastings algorithm with the proposal distribution given by a $N_d(\gamma + (h/2)\nabla(\gamma), hI_d)$ where $h > 0$ is a user-specified proposal variance. The use of the gradient in the proposal distribution may lead to much better convergence properties when compared to the standard alternative of a random walk Metropolis algorithm (see Christensen et al. (2000) and Christensen and Waagepetersen (2001)).

A truncated version of the Langevin-Hastings algorithm is obtained by replacing the gradient $\nabla(\gamma)$ in the proposal distribution by

$$\nabla_{\text{trun}}(\gamma) = -\gamma + \left(n_i - \min\{H, A_i \exp(\tilde{y}_i)\}\right)_{i \in I_{\text{ext}}} Q^{\mathsf{T}} \qquad (3.21)$$

where $H > 0$ is a user-specified parameter which can, for example, be taken to be twice the maximal n_i, $i \in I$. As shown in Møller et al. (1998) the *truncated Langevin-Hastings algorithm* is geometrically ergodic.

Benes et al. (2001) consider a fully Bayesian approach for an LGCP where the truncated Langevin-Hastings algorithm is extended with updates of the model parameters. Benes et al. (2001) also discuss convergence of the posterior for the discretized LGCP when the cell sizes tend to zero.

Example 10 Let W be the union of the five frames in the lower right corner in Figure 3.2, let B be the smallest rectangle containing these five frames, and define the discretized Gaussian field $\tilde{Y}$ on a 60×40 rectangular grid I on B. The left plot in Figure 3.6 shows a prediction of $\tilde{Z} = \exp(\tilde{Y})$ given by the conditional mean $E(\tilde{Z}|x)$ of $\tilde{Z}$ using the parameter estimates from Example 8. The minimum and maximum values of $E(\tilde{Z}|x)$ are 0.01 and 0.03, respectively.

3.7.3 Conditional Simulation for Shot-noise G Cox Processes

Conditional simulation for an SNGCP with $\alpha < 0$ follows much the same way as in Section 3.7.1, so we let $0 \le \alpha < 1$ in the sequel. In applications involving MCMC we consider the approximation (3.17) where D is typically equal to an extended region $B_{\text{ext}} \supset B$ as in Section 3.7.1. Let $\psi_j = q(\gamma_j)$, $j = 1, \dots, J$. By (3.17), conditional simulation of Z^J is given by conditional simulation of the vector $(c_1, \psi_1, \dots, c_J, \psi_J)$. Assuming that the measure κ has a density k_κ with respect to Lebesgue measure, $(c_1, \psi_1, \dots, c_J, \psi_J)$ has

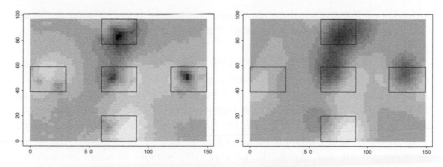

Figure 3.6 *Left: conditional mean of $\tilde{Z}$ under an LGCP model. Right: conditional mean of Z_B^J under an SNGCP model. The same gray scales are used in the two plots. The five observation frames are those found in the lower left plot in Figure 3.2 but here rotated back $90°$.*

density proportional to $\exp(-\psi_J) \prod_{i=1}^{J} k_\kappa(c_i)$ for $(c_1, \ldots, c_J) \in D^J$ and $0 < \psi_1 < \ldots < \psi_J$. Note that the conditional density

$$p(x|c_1, \psi_1, \ldots, c_J, \psi_J) \propto \exp\left(-\int_W Z^J(x)\mathrm{d}x\right) \prod_{j=1}^{n} Z^J(x_j)$$

does not depend on the order of (c_j, ψ_j), $j = 1, \ldots, J$. So we can forget the constraint $0 < \psi_1 < \ldots < \psi_J$, and simply consider the posterior obtained with the prior density proportional to $\exp(-\psi_{\max}) \prod_{i=1}^{J} k_\kappa(c_i)$ where $\psi_{\max} = \max\{\psi_1, \ldots, \psi_J\}$. The density of $(c_1, \psi_1, \ldots, c_J, \psi_J)$ given x is then proportional to

$$\left[\exp\left(-\int_W Z^J(x)\mathrm{d}x\right) \prod_{j=1}^{n} Z^J(x_j)\right]\left[\exp(-\psi_{\max}) \prod_{i=1}^{J} k_\kappa(c_i)\right].$$

For the special case of shot-noise Poisson-gamma models, Wolpert and Ickstadt (1998) use a certain data augmentation scheme and a Gibbs sampler, but we consider a more simple approach where we just apply a standard random scan single-site Metropolis algorithm. A proposal for our Metropolis algorithm is obtained by picking a j in $\{1, \ldots, J\}$ uniformly at random, and then replacing (c_j, ψ_j) by (c'_j, ψ'_j) generated from the uniform distribution on $B_{\mathrm{ext}} \times]c_j - s_c, c_j + s_c[$, where $s_c > 0$ is a user-specified parameter. In the implementation of the algorithm it is helpful to introduce an auxiliary variable U which takes the value j if $\psi_{\max} = \psi_j$, $j = 1, \ldots, J$, and thus keeps track of the maximal ψ_j. The variable U can further be used to monitor convergence for the Markov chain, since U follows the uniform distribution on $\{1, \ldots, J\}$ when the chain is in equilibrium.

The algorithm can straightforwardly be extended to accommodate a fully Bayesian analysis with priors also on the model parameters. Fully Bayesian analysis of Poisson-gamma shot-noise processes is considered in Wolpert and Ickstadt (1998) and Best et al. (2000), for example.

Example 11 Let W and $B = [0, 150] \times [0, 100]$ be as in Example 10, let $B_{ext} = [-40, 190] \times [-40, 140]$, and consider the fitted SNGCP from Example 9. The right plot in Figure 3.6 shows the posterior mean of $(Z^J(\boldsymbol{x}))_{\boldsymbol{x} \in B}$ obtained with $J = 2000$, using a sample obtained by subsampling each 10^4 state of a Markov chain of length 10^7 generated by the Metropolis algorithm. The minimal and maximal values of $E(Z^J(\boldsymbol{x})|x)$, $\boldsymbol{x} \in B$, are the same as in Example 10, but the predicted intensity surface is more smooth under the SNGCP.

3.8 Discussion

Remark 1 (Likelihood-based inference) Concerning parameter estimation we have focused on minimum contrast estimation which is advantageous for computational reasons. This is somewhat out of line with modern spatial statistics where likelihood-based methods (either in a Bayesian or frequentist framework) attracts increasing attention. In general one must expect minimum contrast estimation to be less efficient than likelihood-based estimation. The minimum contrast estimates can also be very sensitive to the choice of user-specified parameters.

If we consider LGCPs and SNGCPs with $0 \leq \alpha < 1$, then two problems appear in connection with likelihood-based inference. First, in order to apply MCMC methods, we need to approximate the processes either by discretizing the Gaussian field for an LGCP or by using the truncated sum (3.17). Second, in order to get a good approximation, we need to use either a fine grid I or a large truncation value J. Conditional simulation for the approximate processes thereby becomes computationally very demanding. Also the storage of samples for Monte Carlo estimation of the likelihood may be problematic. For Neyman-Scott processes and SNGCPs with $\alpha < 0$ and a moderate value of κ, the conditional distribution will in contrast typically be of moderate dimension, and there is no need to introduce an approximation (apart from possible edge effects when the kernel density f does not have bounded support).

Remark 2 (Definition of an SNGCP) In Section 3.5.3 we gave a direct definition of an SNGCP. Below we briefly discuss how such processes have been introduced in Brix (1999).

Brix (1999) defines first a G measure M. This is a random measure so that $M(A_1), \ldots, M(A_n)$ are independent for disjoint bounded regions $A_1, \ldots, A_n$, and each $M(A_i)$ follows a so-called G distribution which is parameterized in a certain way by $(\alpha, \kappa(A_i), \tau)$. It turns out that a G measure is a locally finite measure of the form $M(A) = \sum_j \gamma_j \mathbf{1}\{\boldsymbol{c}_j \in A\}$

where $\{(c_j, \gamma_j)\}$ is defined as in Section 3.5.3. Secondly, since M is purely atomic, Brix views a G Cox process as a random measure Φ: conditionally on M, we have that Φ is a random measure so that $\Phi(A_1), \ldots, \Phi(A_n)$ are independent for disjoint regions $A_1, \ldots, A_n$, and each $\Phi(A_i)$ is Poisson distributed with mean $M(A_i)$. Finally, in order to obtain some smoothing, Brix defines an SNGCP by extending the definition of M to $M(A) = \sum_j \gamma_j K(A, c_j)$ where each $K(\cdot, c_j)$ is an arbitrary probability measure. When $K(\cdot, c_j)$ is given by a density function $k(\cdot, c_j)$ as in Section 3.5.3, we obtain that Φ is now a nonatomic measure, which can be identified with our SNGCP X given by (3.14).

G Cox processes have nice properties under aggregation due to the simple form of the distribution of aggregated counts $\Phi(A_1), \ldots, \Phi(A_n)$. For example under a Poisson-gamma model (Wolpert and Ickstadt 1998), the counts are negative binomial distributed. However, one should note that these simple properties are not valid for SNGCPs, due to the smoothing by the kernel $k(\cdot, \cdot)$. See also the discussion in Richardson (2001) and Møller (2001a).

Remark 3 (Extension of SNGCPs) One may note that both SNGCPs and Neyman-Scott processes are special cases of Cox processes driven by a random intensity surface of the form

$$Z(\boldsymbol{x}) = \sum_j \lambda_j k(x, c_j) \tag{3.22}$$

where $\{(c_j, \lambda_j)\}$ is a Poisson process with an intensity measure of the form $\nu(\mathrm{d}(c, \lambda)) = \kappa(\mathrm{d}c)\zeta(\mathrm{d}\lambda)$ (i.e. a product measure). Here ν is assumed to be locally finite, but the marginal process $\{c_j\}$ may not be locally finite, since it is possible that $\zeta([0, \infty)) = \infty$, cf. the case of an SNGCP with $0 \leq \alpha \leq 1$.

Many properties for Neyman-Scott processes and SNGCPs follow from general results for the extended model. For example, if $k(\boldsymbol{x}, c) = k_0(\boldsymbol{x} - c)$ and $\kappa(\cdot) = \kappa_0 |\cdot|$ where $\kappa_0 > 0$ is a parameter, the process is stationary with intensity $\rho = \kappa_0 \int \lambda \zeta(\mathrm{d}\lambda)$. Assuming $\int \lambda^2 \zeta(\mathrm{d}\lambda) < \infty$, the pair correlation exists and is given by

$$g_0(\boldsymbol{x}) = 1 + (\kappa_0/\rho)^2 \int \lambda^2 \zeta(\mathrm{d}\lambda) \int k_0(\boldsymbol{x}_0) k_0(\boldsymbol{x} + \boldsymbol{x}_0) \mathrm{d}\boldsymbol{x}_0.$$

Equations (3.7) and (3.16) are special cases of this result.

As in Møller (2001a) we suggest that more attention should be drawn to "modified" Neyman-Scott models of the type (3.22) with $\{c_j\}$ a locally finite Poisson process. One can thereby utilize the additional flexibility offered by the random coefficients λ_j and still work with a locally finite conditional distribution in connection with Bayesian inference or maximum likelihood estimation.

Remark 4 (Comparison of models) The focus in this chapter has been on inferential and computational aspects of various parametric models for Cox processes. The advantages and disadvantages of LGCPs, SNGCPs, and the Heikkinen and Arjas (1998) model for inhomogeneous Poisson processes are also discussed in Richardson (2001) and Møller (2001a). Which model is most appropriate depends of course on prior knowledge and the purpose of the statistical analysis. LGCPs provide easily interpretable Cox process models with some flexibility in modelling aggregated point patterns when the aggregation is due to an environmental heterogeneity. Even more flexibility may be obtained by using SNGCPs and their extensions (Remark 3), and such models seem natural when the aggregation is due to clustering around a process of cluster centres. They seem also appropriate for nonparametric Bayesian modelling (Wolpert and Ickstadt 1998, Richardson 2001, Møller 2001a).

CHAPTER 4

Extrapolating and Interpolating Spatial Patterns

M.N.M. van Lieshout   *A.J. Baddeley*

4.1 Introduction

Observations of a spatial pattern are typically confined to a bounded region of space, while the original pattern of interest can often be imagined to extend outside. Much attention has been paid to statistical inference for models of the pattern given only the partial observations in the sampling window. Less attention has been given to *prediction* or *extrapolation* of the process (i.e. of the same realisation of the process) beyond the window of observation, conditional on the partially observed realisation. A motivating example is the charting of geological faults encountered during coal mining (Baddeley et al. 1993, Chilès 1989). It is of interest to predict the likely presence of geological faults outside the region mined so far, and thereby to choose between various mining strategies. Other examples may be found in image processing, for instance the problem of replicating a texture beyond the region where it has been observed as in the editing of a video image so that a foreground object is removed and replaced seamlessly by the background texture (De Bonet 1997). Partial observation of a spatial pattern may also include effects such as aggregation by administrative regions, deletion of part of the pattern, and the unobservability of a related pattern. Recovery of full information in this context might be called *interpolation*; it resembles a missing data problem. In the mining problem discussed above, mapped charts represent only those parts of geological faults which were physically encountered. Gaps may arise because the mined region is not convex both at its outer boundary and within this boundary, because pillars of unmined material remain. Hence it is of interest to join observed line segments together and to interpret them as part of the same continuous fault zone, a process that is known as 'interpretation' by geologists. As another example, geostatistics deals with predicting values of a spatial random process (e.g. precipitation or pollution measurements) from observations at known locations (e.g. Journel and Huijbregts 1978, Cressie 1993, Stein 1999) and interpolation techniques have been developed under the name of conditional simulation for Gaussian and other second-order random fields, as well as for discrete Markov random field models. Relatively few conditional simulation techniques have been developed for spatial processes of geometric features such as points, line segments and filled shapes. Those that exist are based largely on Poisson processes and the associated Boolean models (Lantuéjoul 1997, Kendall and Thönnes 1999, van Lieshout and van Zwet 2001). A major obstacle is the scarcity of spatial models that are both realistic and tractable for simulation. Some exceptions are the following. There has been much interest in the conditional simulation of oil-bearing reservoirs given data obtained from one or more exploration wells (Haldorsen 1983, Chessa 1995). The wells are essentially linear transects of the spatial pattern of reservoir sand bodies. Typically the sand bodies are idealised as rectangles with horizontal and vertical sides of independent random lengths, placed at random locations following a Poisson point process. For line segment processes, Chilès (1989) presents some stochastic models with particular application to modelling geological faults (based largely on Poisson processes), geostatistical inference, and possibilities for conditional simulation; Hjort and Omre (1994) describe a pairwise interaction point process model for swarms of small faults in a fault zone, and Stoica et al. (2000,2001) study a line segment process for extracting linear networks from remote sensing images. Some of these authors have correctly noted the sampling bias effect attendant on observing a spatial pattern of geometric features within a bounded window (analogous to the 'bus paradox'). Techniques from stochastic geometry need to be enlisted to check the validity of simulation algorithms. Extrapolation or interpolation of a spatial pattern entails fitting a stochastic model to the observed data, and computing properties of the conditional distribution of this model given the observed data. We will discuss a variety of stochastic models for patterns of geometric objects, and treat typical issues such as edge effects, occlusion and prediction in some generality. Subsequently, we shall focus on the problem of identifying clusters in a spatial point pattern, which can be regarded as interpolation of a two-type point pattern from observations of points of one type only, the points of the other type being the cluster centres (Baddeley and van Lieshout 1993, Lawson 1993b, van Lieshout 1995, van Lieshout and Baddeley 1995). Applications may be found in epidemiology, forestry, archaeology, coal mining, animal territory research, and the removal of land mines.

4.2 Formulation and Notation

In this section we describe the general framework considered throughout. The spatial pattern is a random closed set (Matheron 1975, Stoyan et al. 1995) U in $\mathbb{R}^d$, typically $d = 2$ or 3. The distribution of U is governed by a parameter θ in some space Θ.

4.2.1 Germ–grain Models

All models considered in this chapter are *germ–grain models* (Stoyan et al. 1995) constructed as follows. There is an underlying process $X = \{X_i, \ i = 1, 2, \ldots\}$ of germs in $\mathbb{R}^d$, each associated with a random compact set Z_i (the 'grain') in $\mathbb{R}^d$ specified by a parameter in some space $\mathcal{Z}$. The 'complete data' process $W = \{(X_i, Z_i)\}$ consists of pairs of germs with their associated grains and hence can be seen as a marked point process. The union of the translated grains, $U = \bigcup_i(X_i + Z_i)$, forms the germ-grain model. We shall be concerned mostly with spatial cluster processes, which can be formulated as germ-grain models where the X_i are the cluster centres, Z_i is the cluster of points or objects associated with centre X_i translated back to the origin (i.e. Z_i is a random finite set of geometric objects), and U is the union pattern. We will sometimes refer to the X_i as the *parents* and to $X_i + Z_i$ as the *daughters* or *offspring* of X_i. If both the cluster centres and their offspring are points, $\mathcal{Z}$ is the space $\mathcal{N}$ consisting of all finite point patterns in $\mathbb{R}^d$. The complete data W then consists of the patterns X and U together with information mapping each member of U to its cluster centre in X. Note that if $X = \{X_i, i = 1, 2, \ldots\}$ is a homogeneous Poisson point process, and the Z_i are i.i.d. the random closed set U is a *Boolean model* (see pp. 484–502 Serra 1982). The common distribution of Z_i is called the distribution of the *typical grain*; the germs X_i play only an indirect role. In practice, one observes the intersection $Y = U \cap A$ of U with a compact window $A \subseteq \mathbb{R}^d$. Mostly the window A is fixed and known. More generally, one may assume that A is an observable random set and condition on it, effectively implying A should be ancillary for θ and independent of U. The requirement that A be observable excludes, for example, random thinning models (Cressie 1993, Stoyan et al. 1995). These are unidentifiable in the sense that one cannot distinguish between a point process of low intensity and a heavily thinned point process of higher intensity, without imposing further assumptions.

4.2.2 Problem Statement

The goal is, given data $\mathbf{y} = U \cap A$, to obtain estimates of the conditional expectations of random variables associated with U or W. Note that in the latter case, W will contain grains Z_i^* such that $X_i^* + Z_i^*$ hits the boundary of A. Hence, any extrapolation technique will have to extend Z_i^* as well as locate germ-grain pairs not hitting A. It is important to realise that the individual objects $X_i + Z_i$ in the germ-grain model are not assumed to be observable separately. They are merely an intermediate stage in the construction of the model for the random set U. For example, any object $X_i + Z_i$ which is completely occluded, i.e. contained in the union of other objects, is not observable and may as well be absent. Consequently our analysis must depend only on the union set U and not on the representation of U as a union of objects $X_i + Z_i$. In other words, if the data image $\mathbf{y}$ can be represented in two different ways

$$\mathbf{y} = \bigcup_{i=1}^{n}(X_i + Z_i) \cap A = \bigcup_{j=1}^{m}(X_j' + Z_j') \cap A$$

then inference based on either representation must yield identical results. This rules out mark-correlation techniques (Penttinen and Stoyan 1989). Specialising to spatial cluster analysis, inference focuses on the conditional expected number of clusters, the conditional mean number of points per cluster and the posterior distribution of centre locations as well as the strength of evidence for clustering. As for occlusion effects, the whole essence of the problem is that we do not know which data points belong to the same cluster. Below, we adopt a Bayesian strategy and base inference on the posterior distribution of W given $\mathbf{y}$. The parameter vector θ will be estimated by Monte Carlo maximum likelihood (Gelfand and Carlin 1993, Geyer 1999).

4.2.3 Edge Effects and Sampling Bias

Edge effects and sampling bias are bound to arise when a spatial pattern of unbounded extent is observed in a bounded frame (Baddeley 1999). In this section, we illustrate these problems for partial realisations of a Poisson process of geometric objects. Although the Poisson assumption allows for explicit computations, the essential complexities of the general problem are already present. Thus, assume that the germ process $X = \{X_i\}$ is a homogeneous Poisson point process in $\mathbb{R}^d$ with intensity $\lambda > 0$, that the grains Z_i are i.i.d. random compact sets, and that $A \subseteq \mathbb{R}^d$ is a fixed, compact window. We wish to generate a realisation of $U \cap A$. The approach taken will be to sample those objects which wholly or partly intersect A, and to clip the resulting pattern to the window A. First, note that a translated grain $X_i + Z_i$ intersects A if and only if $X_i \in A \oplus \check{Z}_i$, where $A \oplus B = \{a + b : a \in A,\, b \in B\}$ is the Minkowski sum of two sets $A, B \subseteq \mathbb{R}^d$ and $\check{A} = \{-a : a \in A\}$ is the reflection of A about the origin (Matheron 1975, Serra 1982, Stoyan et al. 1995). Hence, the germ–grain pairs (X_i, Z_i) for which $X_i + Z_i$ hits A form an inhomogeneous Poisson process whose intensity measure has density $\lambda \mathbf{1}\{x \in A \oplus \check{Z}\}$ with respect to the product of Lebesgue measure and the probability distribution of the grains. Write $|\cdot|$ for d-dimensional volume. Then the number of objects intersecting A is Poisson distributed with mean

$$\lambda E(|A \oplus \check{Z}|) \tag{4.1}$$

where the expectation is with respect to the distribution of the typical grain Z, provided (4.1) is finite. Given n objects are present, they are i.i.d.

with density $1\{x \in A \oplus \check{Z}\}/E(|A \oplus \check{Z}|)$. Turning to the marginal grain distribution, it should be noted that the grains Z_i corresponding to objects which intersect A are not a random sample from the distribution of the typical grain Z. Instead, their distribution is weighted in the sense that Z_i are i.i.d. with distribution

$$P_A(Z \in \cdot) = \frac{E\left(1\{Z \in \cdot\} \,|A \oplus \check{Z}|\right)}{E(|A \oplus \check{Z}|)} \tag{4.2}$$

where E denotes the expectation with respect to the distribution of the typical grain Z. Thus the sampling bias favours larger grains: a larger object is more likely than a smaller object to intersect A. The sampling bias also depends on the geometry and relative orientations of A and Z. For further information, see Serra (1982) or Stoyan et al. (1995). To simulate $U \cap A$, the properties just described can be used if the function $f(Z) = |A \oplus \check{Z}|$ and the distribution (4.2) can be evaluated analytically. In two dimensions, if A is a disc of radius r and Z is almost surely convex with nonempty interior, then by Steiner's formula (page 200 Santaló 1976)

$$|A \oplus \check{Z}| = |Z| + r \operatorname{length}(\partial Z) + \pi r^2 \tag{4.3}$$

almost surely, where $\operatorname{length}(\partial Z)$ denotes the length of the boundary of Z. Hence (4.2) is a mixture of the area-weighted, the length-weighted and the unweighted typical grain distribution. Of course, if Z is a cluster of points, it is not convex. However, if the diameter of Z is almost surely bounded by D (say), we can generate centres $X_i \in A \oplus B_D$ where B_D is the disc of diameter D, form the associated Z_i and clip $X_i + Z_i$ to A. Similarly, one can reduce to the case where A is convex, or even a disc, by simply enclosing A in a larger, convex region A^+ such as the convex hull or circumcircle of A, generating a simulated realisation of U in A^+, and clipping it to A.

4.2.4 Extrapolation

When extending a germ-grain model beyond the observation window, two cases may be distinguished, namely

(i) extending grains Z_i^* such that $X_i^* + Z_i^*$ hits the boundary of A based on $U \cap A$;

(ii) extending the pattern U beyond the window A based on $U \cap A$.

Below we discuss several geometric aspects in some generality. Specific aspects related to spatial cluster processes will be treated in subsequent sections. For Poisson germ-grain models, the conditional distribution of $\{X_i^* + Z_i^* : i = 1, 2, \ldots\}$ given $\{(X_i^* + Z_i^*) \cap A : i = 1, 2, \ldots\}$ is such that the $X_i^* + Z_i^*$ are conditionally independent, and the conditional distribution of $X_i^* + Z_i^*$ depends only on $(X_i^* + Z_i^*) \cap A$ (Daley and Vere-Jones 1988, Last 1990, Kingman 1993, Reiss 1993). Note that this conditional distribution as well as the law of $(X_i^* + Z_i^*) \cap A$ may have atoms, as for example if there is a non-zero probability that a single object $X_i^* + Z_i^*$ covers A completely, or, in the conditional case, if a grain is specified fully by its restriction to A. Atoms need to be treated separately using integral-geometric factorisation techniques (Santaló 1976). If the distribution governing X is not that of a Poisson point process, as for spatial clustering problems, the grains can no longer be extended independently of each other. Other obstacles arise from the unobservability of the individual objects in the pattern (cf. section 4.2.2), and we need to extend grains based on the union set $U \cap A$. Sometimes, $U \cap A$ suffices to determine the individual sets $(X_i^* + Z_i^*) \cap A$; more often it will not be possible to determine the components uniquely from U especially if the window A is not convex or if objects may occlude one another. Indeed, the identification of the offspring partitioning is the whole point of spatial clustering. To conclude this section, note that alternative classes of models include the various Poisson-based constructions described in Chilès (1989), chapter XIII in Serra (1982), and Arak-Surgailis-Clifford mosaics and random graphs in Arak et al. (1993). We use germ–grain models mainly because they are quite flexible while remaining relatively simple from a computational point of view: Markov chain Monte Carlo simulation methods are available by combining existing methods for point processes and for Poisson processes of geometric objects, and parametric and nonparametric inferential methods can be carried over from existing methods for spatial point processes. Moreover, in the alternative models listed above, the geometric features may be connected (e.g. several line segments may have a common endpoint) in a fashion which is inappropriate to most of the applications considered here, although positively desirable for other applications such as random tessellations.

4.3 Spatial Cluster Processes

The identification of centres of clustering is of interest in many areas of applications, including archeology (Hodder and Orton 1976), mining (Chilès 1989, Baddeley et al. 1993) and animal territory research (Blackwell 1998). In disease mapping the identification of cluster centres is of interest (Marshall 1991; see also Chapter 14 of this volume) and mine field detection relies on separating clusters of land mines from clutter of other kinds (Dasgupta and Raftery 1998, Cressie and Lawson 2000). Most traditional clustering approaches build a tree based on some similarity measure, for example, Mardia et al. (1979), Chatfield and Collins (1980), Kaufman and Rousseeuw (1990) or other textbooks on multivariate statistics (see also Chapter 1). From this tree, the number of clusters and the corresponding partition are decided in an ad hoc (and mostly subjective) manner. More recently, model based clustering techniques (Dasgupta and Raftery 1998, Deibolt and Robert 1994) consider finite mixture models. The number of groups is determined by a Bayes factors or AIC criterion, and given the number of mixture components, model parameters are estimated by maximum likelihood, often using a variant of the EM algorithm. Most applications also allow a 'do not know' class for outliers or noise. The cluster centres only play an implicit role – approximated by the centre of gravity, principal axis or other 'mean' of the detected clusters – if they appear at all. Notable exceptions are Lund et al. (1999) and Lund and Thönnes (2000) who model uncertainty in point locations by means of a cluster process consisting of at most a single point, and van Lieshout et al. (2001) who employ variational analysis in the space of intensity measures of the parent point process. In contrast, following up on earlier work (Baddeley et al. 1993, van Lieshout 1995, van Lieshout and Baddeley 1995), this chapter advocates the use of point process and germ–grain models (see Section 4.2.1). A virtue of this approach is that the number of clusters, the locations of their centres, and the grouping or labelling of observed points into clusters, are intrinsic aspects of the underlying process (rather than additional parameters) and are all treated simultaneously. The most general model we consider is the independent cluster process introduced in Section 4.3.1, but most attention will be focussed on the computationally convenient Cox cluster processes (Section 4.3.2). The cluster formation densities are derived in Section 4.3.3 below.

4.3.1 Independent Cluster Processes

Let X be a point process on $\mathbb{R}^d$ and associate with each X_i a finite cluster Z_i of points 'centred' at the origin of $\mathbb{R}^d$. Throughout we will assume that the grains Z_i are conditionally independent. The union of offspring $U = \cup_i(X_i + Z_i)$ is an *independent cluster process* (pp. 236–238 in Daley and Vere–Jones 1988, pp. 75–81 and 148 ff. in Cox and Isham 1980). Technical conditions of finiteness and measurability must be satisfied for such a process to exist, see p. 236 in Daley and Vere-Jones (1988). The data consist of a realisation of $Y = U \cap A$ in a compact window $A \subseteq \mathbb{R}^d$ of positive volume. Thus,

$$\mathbf{y} = \{\, y_1, \ldots, y_m \,\}, \qquad m > 0, \quad \mathbf{y} \subseteq A$$

is a configuration of daughters in A. The above formulation is quite flexible, in that it retains the possibility of locating putative cluster parents outside the window A to counteract sampling bias effects (see the discussion in Section 4.2.3) and of grain characteristics such as the daughter intensity or the spread of the cluster to be randomly and spatially varying. In order to be able to base inference on penalised likelihoods, we shall restrict the germ process to lie inside some compact set $\mathcal{X} \subseteq \mathbb{R}^d$ of positive volume, and assume that its distribution is absolutely continuous with respect to a unit rate Poisson point process on $\mathcal{X}$. For each $\xi \in \mathcal{X}$ we are given the distribution Q_ξ of a finite point process Z_ξ on a compact subset $\tilde{\mathcal{X}}$

of $\mathbb{R}^d$; Z_ξ represents the offspring of a parent ξ translated back to the origin to fit in the general germ-grain model of Section 4.2.1. We assume that Q_ξ is absolutely continuous with a density $g(\cdot|\xi)$ with respect to the distribution of a unit rate Poisson process on $\tilde{\mathcal{X}}$. Thus $\mathcal{Z} = \mathcal{N} = \mathcal{N}_{\tilde{\mathcal{X}}}$, the family of finite point configurations in $\tilde{\mathcal{X}}$. To ensure existence of U, we shall assume that the family of densities is jointly measurable seen as a function $g : \mathcal{X} \times \mathcal{N} \to \mathbb{R}^+$. More generally, we could have set $\tilde{\mathcal{X}} = \mathbb{R}^d$ equipped with some finite diffuse intensity measure $\nu(\cdot)$, with the assumption that Q_ξ is absolutely continuous with respect to the distribution of a Poisson process with intensity measure $\nu(\cdot)$. It is of interest to note that when X is a Poisson process and we extend the process onto the whole of $\mathbb{R}^d$, then Q may be almost surely reconstructed from a single joint observation of parents and daughters (Milne 1970, Bendrath 1974). Table 4.1 summarises standard nomenclature for special cases of the independent cluster model (Stoyan et al. 1995, Daley and Vere-Jones 1988).

Parents X	Clusters Z	Name of process U
general	general	Independent cluster process
Poisson	general	Poisson cluster process
general	Poisson	Cox cluster process
Poisson	Poisson	Neyman-Scott process
Poisson (homogeneous)	Poisson (uniform in disc)	Matérn cluster process
Poisson (homogeneous)	Poisson (Gaussian)	Modified Thomas process

Table 4.1 *Standard nomenclature for independent cluster processes.*

4.3.2 Cox Cluster Processes

For simplicity, most attention will be focussed on the Cox cluster process model, where each grain Z_ξ, $\xi \in \mathcal{X}$, is a realisation of an *inhomogeneous Poisson point process* on $\tilde{\mathcal{X}}$ with *intensity function* $h(\cdot + \xi|\xi) : \tilde{\mathcal{X}} \to [0, \infty)$. In other words, a parent point ξ is replaced by a Poisson number of offspring with mean $H(\xi) = \int_{\tilde{\mathcal{X}}} h(t + \xi|\xi)\, dt \in (0, \infty)$, and given the number of offspring their locations are independently and identically distributed with probability density $f(\cdot) = h(\cdot|\xi)/H(\xi)$ on $\xi + \tilde{\mathcal{X}}$ (with respect to Lebesgue measure). We shall assume the intensity function $h(\cdot|\cdot)$ to be jointly measurable in its arguments, as well as integrable so that $H(\xi) < \infty$ for all $\xi \in \mathcal{X}$. As in Dasgupta and Raftery (1998), van Lieshout (1995) and van Lieshout and Baddeley (1995), scatter noise and outliers – also known as *orphans* – are modelled by a Poisson point process of constant intensity $\epsilon > 0$ independently of all Z_ξ. This fits into the germ–grain framework of Section 4.2.1 by introducing an extra dummy or 'ghost' parent x_0. We shall write $h(\cdot|x_0) \equiv \epsilon$, and denote its integral over $\mathcal{X} \oplus \tilde{\mathcal{X}}$ by

$H(x_0)$. By the superposition property of Poisson processes, conditional on $X = \mathbf{x} = \{\, x_1, \ldots, x_n \}$, the combined offspring form a Poisson point process on $\mathcal{X} \oplus \tilde{\mathcal{X}}$ with intensity function

$$\lambda(\cdot \mid \mathbf{x}) = \epsilon + \sum_{i=1}^{n} h(\cdot|x_i) \tag{4.4}$$

with the convention that $h(t|x_i) = 0$ if $t \notin x_i + \tilde{\mathcal{X}}$, $i = 1, \ldots, n$. The marginal distribution of U is that of a Cox point process Stoyan et al. (1995). Often, the intensity function $h(t|\xi)$ will depend only on the distance $d(\xi, t)$ between ξ and t. An example is

$$h(t|\xi) = \begin{cases} \mu & \text{if } d(\xi, t) \leq R_h \\ 0 & \text{otherwise} \end{cases} \tag{4.5}$$

which, if X is also a Poisson process, is known as the *Matérn cluster process* (Matérn 1986). Another interesting special case is (for $d = 2$, say)

$$h(t|\xi) = \frac{\mu}{2\pi\sigma^2} e^{-d(\xi,t)^2/2\sigma^2}. \tag{4.6}$$

According to (4.6), the daughters follow an isotropic Gaussian distribution with centre ξ. Again if X is a Poisson process, the distribution of U is called the *modified Thomas process*. More generally, the spread σ may depend on ξ. For further details, consult Diggle (1983), Daley and Vere-Jones (1988), or Stoyan et al. (1995). For a Cox cluster process, conditional on $X = \mathbf{x} = \{\, x_1, \ldots, x_n \}$ and the number m of offspring, the points are drawn independently from a finite mixture distribution (Hand 1981, Titterington et al. 1985) with $n + 1$ component distributions determined by the x_i and weights

$$p_i = \frac{H(x_i)}{\sum_{i=0}^{n} H(x_i)}, \qquad i = 0, \ldots, n.$$

If the intensity function h is translation invariant in the sense that $h(t + \xi|\xi) = h(t|0)$ for all $\xi \in \mathcal{X}$ – a common assumption in our spatial context – the weights are identical for all parents except the ghost, a rather unnatural restriction in the finite mixture context. Furthermore, the connection with mixture distributions is lost when the clusters are not Poisson processes. To conclude this section, note that some parents may be childless. In particular, if the clusters Z_ξ are Poisson processes, they have a positive probability of being empty. If in a particular application there is no interest in such parents, one could condition each Z_ξ on $\{\, Z_\xi \neq \emptyset \}$, or consider only those parents having at least one daughter.

4.3.3 Cluster Formation Densities

In order to be able to draw inference about parents and cluster membership, we need the (posterior) distribution of $W = \{\, (x_0, Z_0), \ldots, (X_n, Z_n) \}$, i.e.

of parents X_i marked by their associated grain Z_i, $i = 0, \ldots, n$. We will take a Bayesian approach based on

$$
\begin{aligned}
p_{W|U}(\{(x_i, \mathbf{z}_i)\}_{i \leq n} \mid \mathbf{u}) &\propto P(\mathbf{z}_0, \ldots, \mathbf{z}_n \mid x_0, \ldots, x_n, \mathbf{u})\, p_{X|U}(\mathbf{x} \mid \mathbf{u}) \\
&= c(\mathbf{u})\, P(\mathbf{z}_0, \ldots, \mathbf{z}_n \mid x_0, \ldots, x_n, \mathbf{u})\, p_{U|X}(\mathbf{u} \mid \mathbf{x})\, p_X(\mathbf{x}),
\end{aligned}
\tag{4.7}
$$

the posterior density of W with respect to a unit rate Poisson process on $\mathcal{X}$ marked at $\xi \in \mathcal{X}$ by a label in $\mathcal{P}(\mathbf{u} - \xi)$ according to the uniform distribution on the power set of $\mathbf{u}$ translated back to the origin. The term $p_X(\mathbf{x})$ is the prior density for X with respect to the distribution of a unit rate Poisson process on $\mathcal{X}$, and $c(\mathbf{u})$ a normalising constant depending on the 'data' $\mathbf{u}$. If only the cluster centres are of interest, the posterior density of X (with respect to the distribution of a unit rate Poisson process on $\mathcal{X}$) may be used instead:

$$
p_{X|U}(\mathbf{x} \mid \mathbf{u}) = c'(\mathbf{u})\, p_{U|X}(\mathbf{u} \mid \mathbf{x})\, p_X(\mathbf{x}). \tag{4.8}
$$

We will discuss the choice of prior later on and here describe only the 'forward terms' of cluster formation. Firstly, recall from Section 4.3.1 that conditional on $X = \mathbf{x} = \{x_1, \ldots, x_n\}$, the grains $Z_1, \ldots, Z_n$ associated with $x_1, \ldots, x_n$ respectively are independent with distributions that are absolutely continuous with respect to a unit rate Poisson process on $\tilde{\mathcal{X}}$. Thus, the conditional joint density of $(Z_1, \ldots, Z_n)$ equals

$$
\prod_{i=1}^{n} g(\mathbf{z}_i | x_i)
$$

with respect to the n-fold product measure of unit rate Poisson processes on $\tilde{\mathcal{X}}$. The orphans Z_0 are modelled as a Poisson process of rate $\epsilon > 0$. Again conditioning on $X = \mathbf{x}$, the superposition U is absolutely continuous with respect to the distribution of a unit rate Poisson process on $\mathcal{X} \oplus \tilde{\mathcal{X}}$. Its density at $\mathbf{u} = \{u_1, \ldots, u_m\}$ can be found by summing over all partitions in sibling clusters

$$
p_{U|X}(\mathbf{u} \mid \mathbf{x}) = e^{(1-\epsilon)|\mathcal{X} \oplus \tilde{\mathcal{X}}| - n|\tilde{\mathcal{X}}|} \times \tag{4.9}
$$

$$
\sum_{\varphi} \epsilon^{n(\mathbf{u}_{\varphi^{-1}(\{0\})})} \prod_{i=1}^{n} g(\mathbf{u}_{\varphi^{-1}(\{i\})} - x_i | x_i)\, \mathbf{1}\{\mathbf{u}_{\varphi^{-1}(\{i\})} - x_i \subseteq \tilde{\mathcal{X}}\}
$$

where the sum ranges over all possible offspring-to-parent assignment functions $\varphi : \{1, \ldots, m\} \to \{0, \ldots, n\}$, $\mathbf{u}_{\varphi^{-1}(\{i\})} = \{u_j : \varphi(j) = i\}$ consists of those u_j ascribed to parent x_i by φ, and $n(\cdot)$ denotes cardinality. Equation (4.9) is most readily derived using Janossy densities (page 122 Daley and Vere-Jones 1988). The details can be found in lemma 23 of van Lieshout (1995). Note that (4.9) can be expressed as

$$
e^{(1-n-\epsilon)|\mathcal{X} \oplus \tilde{\mathcal{X}}|} \sum_{\varphi} \epsilon^{n(\mathbf{u}_{\varphi^{-1}(\{0\})})} \prod_{i=1}^{n} g'(\mathbf{u}_{\varphi^{-1}(\{i\})} - x_i | x_i)
$$

where $g'(\cdot - x_i | x_i) = e^{|\mathcal{X} \oplus \tilde{\mathcal{X}}| - |\tilde{\mathcal{X}}|} g(\cdot - x_i | x_i) \mathbf{1}\{\cdot - x_i \subseteq \tilde{\mathcal{X}}\}$ is a density of the translated typical grain with respect to a unit rate Poisson process on $\mathcal{X} \oplus \tilde{\mathcal{X}}$. Next, consider the conditional distribution of the complete model given the cluster centres $x_1, \ldots, x_n$. Since we already derived the conditional joint density of $(Z_1, \ldots, Z_n)$, an identification

$$(\mathcal{Z}^n, \pi_{\mathcal{Z}}^n, \mathcal{A}) \leftrightarrow (\mathcal{N}_{\mathcal{X} \times \mathcal{Z}}, \xi_{\mathbf{x}}, \mathcal{B})$$

of grain vectors $(\mathbf{z}_1, \ldots, \mathbf{z}_n) \in \mathcal{Z}^n$ with the marked point configuration $\{(x_1, \mathbf{z}_1), \ldots, (x_n, \mathbf{z}_n)\} \in \mathcal{N}_{\mathcal{X} \times \mathcal{Z}}$ is needed. Here $\mathcal{Z} = \mathcal{N}_{\tilde{\mathcal{X}}}$ is the grain space (cf. Section 4.2.1) consisting of all finite point configurations, $\pi_{\mathcal{Z}}^n$ is the n-fold product measure of unit rate Poisson processes on $\tilde{\mathcal{X}}$, $\mathcal{A}$ is the usual Borel product σ-algebra of the weak topology (Daley and Vere-Jones 1988), and $\mathcal{B}$ the Borel σ-algebra of the weak topology on marked point patterns. To do so, define a measurable bijection $i_{\mathbf{x}}$ (in the sense that the complement of the range of $i_{\mathbf{x}}$ has measure zero under $\xi_{\mathbf{x}}$) depending on the parent pattern $\mathbf{x} = \{x_1, \ldots, x_n\}$ by

$$i_{\mathbf{x}} : (\mathbf{z}_1, \ldots, \mathbf{z}_n) \mapsto \{(x_1, \mathbf{z}_1), \ldots (x_n, \mathbf{z}_n)\}.$$

Using the identification thus defined, the measure $\xi_{\mathbf{x}}$ is given by $\xi_{\mathbf{x}}(B) = \pi_{\mathcal{Z}}^n(i_{\mathbf{x}}^{-1}(B))$ for all $B \in \mathcal{B}$. Finally, the conditional distribution of W or equivalently the marks Z_i given (X, U) is discrete, with probabilities

$$P(\mathbf{z}_0, \ldots \mathbf{z}_n \mid x_0, \ldots, x_n, \mathbf{u}) = \qquad\qquad (4.10)$$

$$\frac{\epsilon^{n(\mathbf{z}_0)} \prod_{i=1}^n g(\mathbf{z}_i | x_i)}{\sum_{\varphi} \epsilon^{n(\mathbf{u}_{\varphi^{-1}(\{0\})})} \prod_{i=1}^n g(\mathbf{u}_{\varphi^{-1}(\{i\})} - x_i | x_i) \mathbf{1}\{\mathbf{u}_{\varphi^{-1}(\{i\})} - x_i \subseteq \tilde{\mathcal{X}}\}}$$

provided the union $\cup_i (x_i + \mathbf{z}_i)$ equals $\mathbf{u}$. If $g(\cdot - \xi | \xi)$ is hereditary (cf. Section 4.4.1) for each $\xi \in \mathcal{X}$, the sum in the denominator of (4.10) over all functions φ ascribing parents to each offspring, is non-zero. Otherwise, we have to impose the condition that the grain partition and (X, U) are compatible, in the sense that there exists at least one φ for which the term $\epsilon^{n(\mathbf{u}_{\varphi^{-1}(\{0\})})} \prod_{i=1}^n g(\mathbf{u}_{\varphi^{-1}(\{i\})} - x_i | x_i) \mathbf{1}\{\mathbf{u}_{\varphi^{-1}(\{i\})} - x_i \subseteq \tilde{\mathcal{X}}\}$ is strictly positive (Van Lieshout 1995, theorem 29). For Cox cluster processes, the formulae (4.7)–(4.10) can be greatly simplified. Since Z_{ξ} has density

$$g(\mathbf{z} | \xi) = \exp\left[\int_{\tilde{\mathcal{X}}} (1 - h(t + \xi | \xi)) \, dt\right] \prod_{z \in \mathbf{z}} h(z + \xi | \xi)$$

with respect to a unit rate Poisson process on $\tilde{\mathcal{X}}$, (4.9) reduces to

$$p_{U|X}(\mathbf{u} \mid \mathbf{x}) = \exp\left[\int_{\mathcal{X} \oplus \tilde{\mathcal{X}}} (1 - \lambda(t \mid \mathbf{x})) \, dt\right] \sum_{\varphi} \prod_{i=0}^n \prod_{t \in \mathbf{u}_{\varphi^{-1}(\{i\})}} h(t | x_i)$$

$$= \exp\left[\int_{\mathcal{X} \oplus \tilde{\mathcal{X}}} (1 - \lambda(t \mid \mathbf{x})) \, dt\right] \prod_{j=1}^m \lambda(u_j \mid \mathbf{x}) \qquad (4.11)$$

coding $h(\cdot|x_0) \equiv \epsilon$ for the dummy parent x_0. Thus, (4.11) is in accordance with the fact that the independent superposition of Poisson processes is again a Poisson process, here with intensity $\lambda(\cdot \mid \mathbf{x})$ (cf. (4.4) and the discussion in Section 4.3.2). As for the offspring labelling, (4.10) for a Cox cluster process equals

$$P(\mathbf{z}_0, \ldots, \mathbf{z}_n \mid x_0, \ldots, x_n, \mathbf{u}) = \frac{\prod_{i=0}^{n} \prod_{t \in \mathbf{z}_i} h(t + x_i | x_i)}{\prod_{j=1}^{m} \lambda(u_j \mid \mathbf{x})} \qquad (4.12)$$

whenever $\cup_i (x_i + \mathbf{z}_i) = \mathbf{u}$, see van Lieshout and Baddeley (1995) or corollary 30 in van Lieshout (1995). In terms of the label allocation function $\varphi : \{1, \ldots, m\} \mapsto \{0, 1, \ldots, n\}$ allocating each daughter point to its parent, equation (4.12) implies that the daughters are ascribed to a cluster centre x_I *independently* of one another, with probabilities

$$P(\varphi(j) = I) = \frac{h(u_j | x_I)}{\lambda(u_j \mid \mathbf{x})}.$$

The analogue of this result for finite mixtures with m and n fixed was called the *Random Imputation Principle* by Deibolt and Robert (1994). It was taken as an assumption by Binder (1978) (see page 32). Note the statement holds only for Cox cluster processes, i.e. when the clusters are Poisson.

4.4 Bayesian Cluster Analysis

From Section 4.2.2, recall that the prime object of spatial cluster analysis is to evaluate conditional expectations of quantities such as the number of clusters and the mean number of points per cluster based on the posterior distribution (4.7) of the complete data W given $\mathbf{y}$. In the previous section, we derived the densities associated with cluster formation. In Section 4.4.1 below, we discuss the prior, and investigate properties of the posterior distribution in Section 4.4.2. Then we turn to the problems of generating realisations of (4.7) by Markov chain Monte Carlo methods, and of estimating the model parameters (Sections 4.4.3–4.4.4). In Section 4.4.5, we propose an adaptive coupling from the past algorithm that yields exact samples from (4.7). Throughout, the redwood data set (Strauss 1975, Ripley 1977) is used as an illustration.

4.4.1 Markov Point Processes

In this section we focus on the prior term $p_X(\mathbf{x})$ in (4.7), which we shall assume to be the density of a Markov point process (Ruelle 1969, Preston 1976, Ripley 1977, Ripley 1988, Baddeley and Møller 1989, van Lieshout 2000). Following is a brief summary of the facts we need. Let X be a point process on a compact subset $\mathcal{X} \subseteq \mathbb{R}^d$ of positive volume, whose distribution is absolutely continuous with respect to a unit rate Poisson process on $\mathcal{X}$, say with density $p_X(\cdot)$. Then X is *Markov* at range R in the sense of Ripley (1977) if the ratio

$$\lambda_X(\xi; \mathbf{x}) = \frac{p_X(\mathbf{x} \cup \{\xi\})}{p_X(\mathbf{x})} \tag{4.13}$$

is well-defined for all $\xi \in \mathcal{X}$ (i.e. $p_X(\mathbf{x} \cup \{\xi\}) = 0$ implies $p_X(\mathbf{x}) = 0$; in this case we will also say that $p_X(\cdot)$ is *hereditary*) and depends only on those $x_i \in \mathbf{x}$ for which $d(x_i, \xi) < R$. More generally, the fixed range dependence may be replaced by an arbitrary symmetric neighbourhood relation $\sim$ (so that (4.13) depends on $x_i \sim \xi$ only). Even more general Markov point processes are considered by Baddeley and Møller (1989), and the Markovianity of spatial cluster processes is studied in Baddeley et al. (1996). A (Markov) point process defined by its density with respect to a unit rate Poisson process is said to be *locally stable* if its conditional intensity (4.13) is well-defined and uniformly bounded in both its arguments. To model patterns in which the points tend to avoid coming too close together, it is convenient to consider *pairwise-interaction* processes with densities of the form

$$p_X(\mathbf{x}) = \alpha \prod_{x \in \mathbf{x}} \beta(x) \prod_{x \sim x' \in \mathbf{x}} \gamma(x, x') \tag{4.14}$$

where $\beta : \mathcal{X} \to [0, \infty)$ (the 'intrinsic activity') and $\gamma : \mathcal{X} \times \mathcal{X} \to [0, \infty)$ (the 'pairwise interaction') are measurable functions, γ is symmetric, and $\alpha > 0$ is the normalising constant. This model is well-defined (i.e. the density is integrable) at least whenever $\beta(\cdot)$ is uniformly bounded and $\gamma(\cdot, \cdot) \leq 1$. A standard example of (4.14) is the *Strauss process* (Strauss 1975) with $\beta(\cdot) \equiv \beta > 0$ and

$$\gamma(x, x') = \begin{cases} \gamma & \text{if } d(x, x') \leq R \\ 1 & \text{otherwise} \end{cases} \tag{4.15}$$

where $0 \leq \gamma \leq 1$, which has density

$$p_X(\mathbf{x}) = \alpha \beta^{n(\mathbf{x})} \gamma^{s(\mathbf{x})}$$

where $n(\mathbf{x})$ is the number of points in $\mathbf{x}$ and $s(\mathbf{x})$ is the number of pairs x, x' with $d(x, x') \leq R$. The model favours realisations $\mathbf{x}$ that tend to have more points at distances larger than R than under the Poisson model, that is there is *repulsion* between the points. The special case $\gamma = 0$ in which no R-close point pairs are permitted is known as the *hard core process*; $\gamma = 1$ corresponds to a Poisson process with intensity β. More formally, a point process density $p_X(\cdot)$ is called *anti-monotone* (or repulsive) if

$$\lambda_X(\xi; \mathbf{x}') \leq \lambda_X(\xi; \mathbf{x})$$

for all ξ whenever $\mathbf{x} \subseteq \mathbf{x}'$ and *monotone* (or attractive) if its conditional intensity satisfies

$$\lambda_X(\xi; \mathbf{x}') \geq \lambda_X(\xi; \mathbf{x}).$$

The reader may verify that the Strauss process is repulsive for all $\gamma \leq 1$.

4.4.2 Sampling Bias for Independent Cluster Processes

Note that the restriction Y of an independent cluster process U to some compact observation window A is itself an independent cluster process. Indeed,

$$Y = U \cap A = \bigcup_{x \in X} (x + Z_x) \cap A = \bigcup_{x \in X} (x + (Z_x \cap (A - x))).$$

The distribution $Q_{\xi,A}$ of the grain $Z_\xi \cap (A - \xi)$ associated with ξ in the A-clipped process has density

$$g_A(\mathbf{z}|\xi) = \sum_{k=0}^{\infty} \frac{1}{k!} \int \cdots \int_{(\tilde{\mathcal{X}} \setminus (A-\xi))^k} g(\mathbf{z} \cup \{ v_1, \ldots, v_k\} |\xi) \, dv_1 \ldots dv_k \quad (4.16)$$

with respect to a unit rate Poisson process on $\tilde{\mathcal{X}}$. It follows that the posterior distribution of X given Y is analogous to (4.8), except for the fact that $g_A(\cdot|\cdot)$ features instead of $g(\cdot|\cdot)$. As before, a ghost parent is added to account for scatter noise. For Cox cluster processes, (4.16) simplifies to

$$g_A(\mathbf{z}|\xi) = \exp\left[\int_{\tilde{X}} (1 - h(t + \xi|\xi) \, \mathbf{1}\{t \in A - \xi\}) \, dt \right] \prod_{z \in \mathbf{z}} h(z + \xi|\xi)$$

for $\mathbf{z} \subseteq A - \xi$, the density of a Poisson point process with intensity function $h(\cdot + \xi|\xi) \, \mathbf{1}\{\cdot \in A - \xi\}$. Hence, conditionally on $X = \mathbf{x} = \{x_1, \ldots, x_n\}$, Y is an inhomogeneous Poisson process on A with intensity function

$$\lambda(a \mid \mathbf{x}) = \epsilon + \sum_{i=1}^{n} h(a|x_i), \qquad a \in A,$$

where $\epsilon > 0$ is the background clutter term (cf. Section 4.3.2). As for the prior, one could simply assume the parents to be distributed as a Poisson point process, but it seems more natural to incorporate repulsion at short range to avoid 'over fitting' in the sense of many close parents. Thus, one might take as prior for example a hard core process (cf. Section 4.4.1) with density

$$p_X(\mathbf{x}) = \begin{cases} \alpha \beta^{n(\mathbf{x})} & \text{if } d(x_i, x_j) > R \text{ for all pairs} \\ 0 & \text{otherwise} \end{cases} \quad (4.17)$$

with respect to a unit rate Poisson process on $\mathcal{X}$. Upon observing $Y = \mathbf{y} = \{y_1, \ldots, y_m\}$, the analogue of (4.8) for the A-clipped process is

$$p_{X|Y}(\mathbf{x} \mid \mathbf{y}) = c(\mathbf{y}) \, p_X(\mathbf{x}) \exp\left[\int_A (1 - \lambda(a \mid \mathbf{x})) \, da \right] \prod_{j=1}^{m} \lambda(y_j \mid \mathbf{x}) \quad (4.18)$$

which has posterior conditional intensity

$$\lambda_{X|Y}(\xi; \mathbf{x} \mid \mathbf{y}) = \lambda_X(\xi; \mathbf{x}) \exp\left[-\int_A h(a|\xi)\, da\right] \prod_{j=1}^{m}\left[1 + \frac{h(y_j|\xi)}{\lambda(y_j \mid \mathbf{x})}\right].$$

(4.19)

If the prior density $p_X(\cdot)$ is that of a repulsive Markov point process, the posterior distribution specified by (4.18) is hereditary and repulsive too. The posterior range of interaction depends on the support of the family $h(\cdot|\xi)$, $\xi \in \mathcal{X}$, of intensity functions. If furthermore the prior density $p_X(\cdot)$ is locally stable with bound λ for its conditional intensity and $h(\cdot|\cdot)$ is uniformly bounded in both its arguments by H, then $\lambda_{X|Y}(\xi; \mathbf{x} \mid \mathbf{y}) \leq \lambda(1 + H/\epsilon)^m$ implying local stability of (4.18).

4.4.3 Spatial Birth-and-Death Processes

In this section, we address the problem of sampling from the posterior distribution of the complete data W given partial observations $Y = U \cap A = \{y_1, \ldots, y_m\}$ of a Cox cluster process U within some compact observation window A. Note that since the offspring allocation labels are discrete and distributed according to (4.12), and by the Poisson assumption any daughters in A^c are conditionally independent of those in A, the problem reduces to sampling from the conditional distribution (4.18) of X given Y. Since direct sampling does not seem feasible, we apply Markov chain Monte Carlo techniques. Perhaps the oldest such technique is based on *spatial birth-and-death processes* (Preston 1977), continuous time Markov processes whose transitions are either births or deaths. The traditional choice (Ripley 1977, Baddeley and Møller 1989, Møller 1989) is to take a birth rate proportional to the posterior conditional intensity and a constant death rate. Under mild non-explosion conditions, this procedure converges to the target distribution and hence yields approximate samples if run for long enough (Preston 1977, Møller 1989). A disadvantage is that the total birth rate is difficult to compute, and the product over data points in (4.19) may be very large. For these reasons, we prefer to work with the alternative birth rate

$$b(\mathbf{x}, \xi) = \lambda_X(\xi; \mathbf{x})\left[1 + \sum_{j=1}^{m} \frac{h(y_j|\xi)}{\epsilon}\right]$$

(4.20)

which is less peaked than the posterior conditional intensity, while retaining the desirable property of placing most new-born points in the vicinity of points of $\mathbf{y}$. In order to satisfy the detailed balance equations

$$p_{X|Y}(\mathbf{x} \mid \mathbf{y})\, b(\mathbf{x}, \xi) = p_{X|Y}(\mathbf{x} \cup \{\xi\} \mid \mathbf{y})\, d(\mathbf{x} \cup \{\xi\}, \xi),$$

the death rate for deleting ξ from configuration $\mathbf{x} \cup \{\xi\}$ is

$$d(\mathbf{x} \cup \{\xi\}, \xi) = \frac{\exp\left[\int_A h(a|\xi)\,da\right]}{\prod_{j=1}^m \left[1 + \frac{h(y_j|\xi)}{\lambda(y_j|\mathbf{x})}\right]} \left[1 + \sum_{j=1}^m \frac{h(y_j \mid \xi)}{\epsilon}\right]. \tag{4.21}$$

Note that for any locally stable prior distribution for which $\lambda_X(\xi; \mathbf{x}) \le \lambda$ uniformly in $\mathbf{x}$ and ξ, and any $h(\cdot|\cdot)$ that is uniformly bounded in both its arguments by H, the total birth rate

$$B(\mathbf{x}) = \int_{\mathcal{X}} b(\mathbf{x}, \xi)\,d\xi \le \lambda \left[|\mathcal{X}| + \frac{1}{\epsilon} \sum_{j=1}^m \int_{\mathcal{X}} h(y_j|\xi)\,d\xi\right] := B$$

is bounded from above by a constant $B \le \lambda|\mathcal{X}|(1 + mH/\epsilon)$ that is easy to evaluate for typical choices of $h(\cdot|\cdot)$ such as (4.5) or (4.6). The total death rate from parent configuration $\mathbf{x}$ satisfies

$$D(\mathbf{x}) = \sum_i d(\mathbf{x}, x_i) \ge n(\mathbf{x})(1 + H/\epsilon)^{-m}.$$

Hence, by the Preston theorem (Preston 1977) (see e.g. Baddeley and Møller 1989, Møller 1989), there exists a unique spatial birth-and-death process with transition rates given by (4.20) and (4.21). It has unique equilibrium distribution $p_{X|Y}(\cdot \mid \mathbf{y})$, to which it converges in distribution from any initial state. From an algorithmic point of view, if the current state is $\mathbf{x}$, after an exponentially distributed sojourn time of rate $B + D(\mathbf{x})$, with probability $D(\mathbf{x})/(B + D(\mathbf{x}))$ a point of $\mathbf{x}$ is deleted according to the distribution $d(\mathbf{x}, x_i)/D(\mathbf{x})$; a birth is proposed with the complementary probability $B/(B + D(\mathbf{x}))$ by sampling a candidate ξ from the mixture density $\frac{\lambda}{B}\left[1 + \sum_{j=1}^m \frac{h(y_j|\xi)}{\epsilon}\right]$, which is then accepted with probability $\lambda_X(\xi; \mathbf{x})/\lambda$.

Example: Redwood Seedlings

Figure 4.1 shows the locations of 62 redwood seedlings in a square of side approximately 23 m. The data were extracted by (Ripley 1977) from a larger data set in Strauss (1975). The K-function (Ripley 1979, Ripley 1981) for these data is given in Ripley (1977) and suggests aggregation. As noted by Strauss this is caused by the presence of stumps known to exist in the plot, but whose position has not been recorded. Previous analyses of this data set include that of Strauss (1975), who fitted a model later criticised by Kelly and Ripley (1976). Ripley (1977) concluded we should reject the Poisson hypothesis and remarked that there appears to be both clustering and inhibition between clusters. Diggle (1983) fitted a Poisson cluster process of Thomas type and reported least squares estimates $(25.6, 0.042)$ for the parent intensity and the standard deviation of daughter–parent distances. A goodness of fit test showed adequate fit, but, from a biological

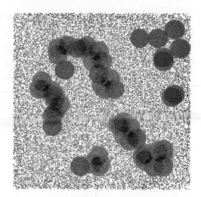

Figure 4.1 *Positions of 62 redwood seedlings in a unit square*

Figure 4.2 *Empirical log posterior parent intensity surface based on a Cox–Matérn cluster process with $R_h = 0.061$ and on average 2.14 points per cluster, noise intensity $\epsilon = 10.0$ and a hard core prior with $R = 0.03$ and $\beta = 1.0$ by spatial birth-and-death over 2.0×10^4 time units. Black corresponds to high values, white to small ones*

point of view, a mean number of 26 stumps seems implausible. In Diggle (1978), a Poisson cluster process of Matérn type was fitted with similar results (radius 0.061 and 29 clusters). None of the above have looked at cluster centre location. This was first studied in Baddeley and van Lieshout (1993) and by Lawson (1993a) who fitted a Poisson–Thomas cluster process and reported 16 parents. An approach based on variational methods can be found in Van Lieshout et al. (2001). In earlier work (Baddeley and van Lieshout 1993, van Lieshout 1995, van Lieshout and Baddeley 1995), we analysed the redwood data using a modified Thomas displacement function (4.6) and a Strauss prior (4.15) with interaction distance 0.084 (Diggle 1983) and $\log \beta = \log \gamma = -10$. Simulation was based on a constant death rate spatial birth-and-death process. Initialising with parameter values $\mu = 7$, $\sigma = 0.042$ and an empty list of cluster centres, we ran the birth-and-death process for 2 time units and found maximum likelihood estimates $\mu = 6.5$ and $\sigma = 0.05$. Here, we use the spatial birth-and-death process with rates (4.20)–(4.21) to sample from the posterior distribution of cluster centres for a Cox model with a Matérn style intensity function given by (4.5) with $R_h = 0.061$ and $\mu = 2.14/(\pi R_h^2)$ as in Diggle (1978), orphan intensity $\epsilon = 10.0$, and a hard core prior with $R = 0.03$ and $\beta = 1.0$. The posterior intensity surface of parents in $\mathcal{X} = [-R_h, 1.0 + R_h]^2$ over 2.0×10^4 time units after a burn-in period of 200.0 units with empty initial state is plotted in Figure 4.2; for the posterior histogram of the number of clusters, see Figure 4.3 (*right*). To indicate the effect of the choice of parameters, the posterior histogram for $\beta = 0.052$ and an average cluster size of 4.3 is shown in Figure 4.3 (*left*). It can be seen that the latter choice shifts the posterior histogram towards fewer cluster centres.

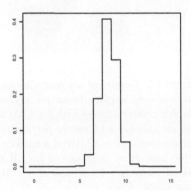 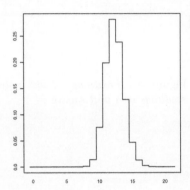

Figure 4.3 *(Left) Posterior histogram for the number of parents given the data of Figure 4.1 based on a Cox–Matérn cluster process with $R_h = 0.061$ and on average 4.3 points per cluster, noise intensity $\epsilon = 10.0$ and a hard core prior with $R = 0.03$ and $\beta = \exp(4.3 + 2.0 \log(\epsilon/(\epsilon + \mu))$ by spatial birth-and-death over 2.0×10^5 time units and (right) for on average 2.14 points and $\beta = 1.0$ over 2.0×10^4 time units as in Figure 4.2.*

4.4.4 Parameter Estimation

In general, the independent cluster model $g(\cdot|\cdot)$ will contain parameters θ that must be estimated. For the Cox cluster model, the parameters are the clutter intensity ϵ as well as parameters of the displacement function $h(\cdot|\cdot)$ specifying the shape, the spread and the number of daughters in each cluster. Moreover, the prior model $p_X(\cdot)$ also contains parameters, but since these are merely used as regularisation to avoid over fitting, we will treat these as fixed. We shall use the Monte Carlo maximum likelihood method for missing data models (Gelfand and Carlin 1993, Geyer 1999). In the context of detecting the centres in an independent cluster process, the observed data consists of a point pattern $\mathbf{y}$, the combined offspring in the window A. The missing data are both the parents and their associated grains. In terms of the cluster formation density derived in Section 4.3.3, the log likelihood ratio with respect to a fixed reference parameter θ_0 can be written as

$$l(\theta) = \log E_{\theta_0} \left[\left. \prod_{i=0}^{n(X)} \frac{g_\theta(Z_i|X_i)}{g_{\theta_0}(Z_i|X_i)} \right| Y = \mathbf{y} \right] \tag{4.22}$$

by importance sampling theory. The Monte Carlo analogue $l_k(\theta)$ of (4.22) is obtained by replacing the expectation by the average in a sample $W_1, \ldots, W_k$ from the complete model under the conditional distribution with parameter θ_0. Differentiating with respect to θ, the parameter of interest, we obtain

$$\nabla l_k(\theta) = \frac{1}{k} \sum_{j=1}^{k} w_{j,\theta_0,\theta} \frac{\nabla \left[\prod_{(X_i,Z_i) \in W_j} g_\theta(Z_i|X_i) \right]}{\prod_{(X_i,Z_i) \in W_j} g_\theta(Z_i|X_i)} \tag{4.23}$$

where

$$w_{j,\theta_0,\theta} = \left(\prod_{(X_i,Z_i) \in W_j} \frac{g_\theta(Z_i|X_i)}{g_{\theta_0}(Z_i|X_i)} \right) \Bigg/ \left(\frac{1}{k} \sum_{j=1}^{k} \prod_{(X_i,Z_i) \in W_j} \frac{g_\theta(Z_i|X_i)}{g_{\theta_0}(Z_i|X_i)} \right)$$

are the importance weights. The well-known EM algorithm (Dempster et al. 1977b) is an iterative procedure based on (4.23) that consists of two steps: the E-step computes the conditional log likelihood given the data and current estimates of the parameters, the M-step maximises the result with respect to the parameter. Thus, the importance weights reduce to 1, but resampling is needed at each step. For a critical evaluation of these and other parameter estimation methods, the reader is referred to Geyer (1999); see also Diggle et al. (1994), Geyer and Møller (1994) and Huang and Ogata (2001). For Cox cluster processes, (4.22) simplifies to

$$l(\theta) = \log E_{\theta_0} \left[\left. e^{\sum_{i=0}^{n(X)} (H_{\theta_0}(X_i) - H_\theta(X_i))} \prod_{i=0}^{n(X)} \prod_{t \in X_i + Z_i} \frac{h_\theta(t|X_i)}{h_{\theta_0}(t|X_i)} \right| Y = \mathbf{y} \right]$$

hence the Monte Carlo score vector (4.23) is

$$\nabla l_k(\theta) = \frac{1}{k} \sum_{j=1}^{k} \left\{ w_{j,\theta_0,\theta} \sum_{(X_i,Z_i) \in W_j} \left[-\nabla H_\theta(X_i) + \sum_{t \in X_i + Z_i} \nabla \log h_\theta(t|X_i) \right] \right\}$$

where as before $\{W_j\}$ is a sample of size k from the conditional distribution of the complete data given $\mathbf{y}$ under the reference parameter θ_0, and $w_{j,\theta_0,\theta}$ are the importance weights.

Example: Cox–Matérn Cluster Process

Consider the Cox–Matérn cluster process on $\mathbb{R}^2$ with offspring governed by (4.5) and independent Poisson background clutter. Treating the range R_h as fixed, the parameter vector is $\theta = (\epsilon, \mu)$. The grain is a finite point process on $\tilde{\mathcal{X}} = B(0, R_h)$, and $H(\xi) = \mu \pi R_h^2$ for each genuine parent $\xi \in \mathcal{X}$. For the dummy parent, $H(x_0) = \epsilon |\mathcal{X} \oplus B(0, R_h)|$. If $\mathcal{X}$ is a convex set, the Steiner formula may be used to find an explicit expression of this area, see Section 4.2.3. By differentiation with respect to the parameter vector, it follows that the components of $\nabla l_k(\theta)$ are the weighted averages of $-|\mathcal{X} \oplus B(0, R_h)| + n(Z_0^j)/\epsilon$ and $-n(W_j) \pi R_h^2 + \sum_{i=1}^{n(W_j)} n(Z_i^j)/\mu$ where $n(W_j)$ denotes the number of genuine parents in W_j, and Z_0^j its orphan cluster. The EM-updates are easily derived:

$$\epsilon^{(n+1)} = \frac{E_{\theta^{(n)}}\left[n(Z_0) \mid Y = \mathbf{y}\right]}{|\mathcal{X} \oplus B(0, R_h)|};$$

$$\mu^{(n+1)} = \frac{E_{\theta^{(n)}}\left[\sum_{i=1}^{n(X)} n(Z_i) \mid Y = \mathbf{y}\right]}{E_{\theta^{(n)}}\left[\pi R_h^2 n(X) \mid Y = \mathbf{y}\right]}.$$

For the example on redwood seedlings (Section 4.4.3) with a unit rate hard core prior at range 0.03 and reference parameter vector (10.0, 183.06) as in Figure 4.2, the Monte Carlo log likelihood ratio for $\epsilon \in (5, 30)$ and $\mu \pi R_h^2 \in (1.14, 7.14)$ is given in Figure 4.4; the solution of the Monte Carlo score equations is $(\hat{\epsilon}_{100}, \hat{\mu}_{100}) = (19.65, 354.15)$. For comparison, the Monte Carlo EM-updates would be $\epsilon = 15.12$ and $\mu = 311.61$ corresponding to 3.64 daughters on average in a cluster.

4.4.5 Adaptive Coupling from the Past

Remarkably, the spatial birth–and–death approach described in Section 4.4.3 can be adapted to yield an *exact* sample from the desired posterior distribution using coupling from the past (Propp and Wilson 1996, Kendall and Møller 2000). Such algorithms are particularly efficient when there is some monotonicity in the state space, and the sampler respects this order. In the context of this chapter, the prior distribution of X is a repulsive Markov point process. Whether the same is true for the posterior distribution depends on the grain distributions Q_ξ. However, for Cox cluster processes, we showed in Section 4.4.2 that the posterior distribution is repulsive and hereditary too. Moreover, (4.20)–(4.21) reverse the inclusion ordering in the sense that if $\mathbf{x} \subseteq \mathbf{x}'$ then $b(\mathbf{x}, \xi) \geq b(\mathbf{x}', \xi)$ for all $\xi \in \mathcal{X}$, while $d(\mathbf{x}, x_i) \leq d(\mathbf{x}', x_i)$ for $x_i \in \mathbf{x}$. Our proof can be found in Section 4.9.3 of van Lieshout (2000). If the displacement functions $h(\cdot|\cdot)$ are uniformly bounded by H, the posterior inherits local stability from the prior. Such properties are particularly pleasing for Bayesian analysis, as they imply

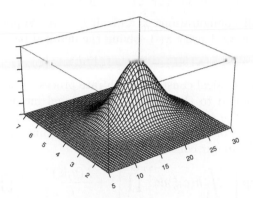

Figure 4.4 *Monte Carlo log likelihood ratio surface as a function of the noise intensity $\epsilon \in (5, 30)$ and the mean cluster size $\mu\pi R_h^2 \in (1.14, 7.14)$ for the redwood seedlings data (Figure 4.1) based on a Cox–Matérn cluster process with $R_h = 0.061$ and reference parameter values such that the average number of points per cluster is 2.14, the noise intensity $\epsilon = 10.0$. We used a hard core prior with $R = 0.03$ and $\beta = 1.0$. One hundred realisations were subsampled from a run of a spatial birth-and-death process over 2.0×10^4 time units after burn-in.*

that the choice of prior is not crucial in these respects. Hence the coupling from the past algorithm of Kendall and Møller (2000) for locally stable point processes in principle applies. Those authors presented their method for the constant death rate dynamics, with a dominating process that is Poisson with an upper bound to the conditional intensity of the distribution to be sampled as intensity. In our context, such a method would be impractical, as in most cases the upper bound will be orders of magnitude too large. For this reason, we present an adaptive coupling from the past algorithm based on (4.20)–(4.21). Suppose a spatial birth-and-death process with transition rates $b(\cdot, \cdot)$ and $d(\cdot, \cdot)$ is available to sample from the posterior density of cluster centres $p_{X|Y}(\cdot \mid \mathbf{y})$, and we have upper and lower bounds

$$b(\mathbf{x}, \xi) \leq \bar{b}(\xi) \tag{4.24}$$

$$d(\mathbf{x} \cup \{\xi\}, \xi) \geq \underline{d}(\mathbf{x} \cup \{\xi\}, \xi) \tag{4.25}$$

holding for all configurations $\mathbf{x}$ and all $\xi \in \mathcal{X}$. Suppose furthermore a unique probability density $\pi(\cdot)$ solving the detailed balance equations

$$\pi(\mathbf{x}) \bar{b}(\xi) = \pi(\mathbf{x} \cup \{\xi\}) \underline{d}(\mathbf{x} \cup \{\xi\}, \xi)$$

exists. For the classical constant death rate process, $\underline{d}(\mathbf{x} \cup \{\xi\}, \xi) \equiv 1$, $\bar{b}(\xi)$ is an upper bound to the posterior conditional intensity at ξ that does not depend on the configuration to which ξ is added, and $\pi(\cdot)$ defines an inhomogeneous Poisson process with intensity function $\bar{b}(\cdot)$. The generic adaptive choice in our context is

$$\bar{b}(\xi) = \lambda \exp\left[-\int_A h(a|\xi) da\right] \prod_{j=1}^{m}\left[1 + \frac{h(y_j|\xi)}{\epsilon}\right] \leq \lambda \prod_{j=1}^{m}\left[1 + \frac{h(y_j|\xi)}{\epsilon}\right]$$

where λ is the prior local stability bound. If a uniform bound is required (Chapter 5), the right hand side above may be replaced by

$$\lambda \sup_{\xi \in \mathcal{X}} \prod_{j=1}^{m}\left[1 + \frac{h(y_j|\xi)}{\epsilon}\right].$$

Similarly, for the transition rates given by (4.20)–(4.21), generic bounds are

$$\bar{b}(\xi) = \lambda\left[1 + \sum_{j=1}^{m} \frac{h(y_j|\xi)}{\epsilon}\right]$$

and

$$\underline{d}(\mathbf{x} \cup \{\xi\}, \xi) = \exp\left[\int_A h(a|\xi)\, da\right]\left[1 + \sum_{j=1}^{m} \frac{h(y_j|\xi)}{\epsilon}\right] / \prod_{j=1}^{m}\left[1 + \frac{h(y_j|\xi)}{\epsilon}\right].$$

The corresponding equilibrium distribution is that of a Poisson process with intensity function $\lambda e^{-\int_A h(a|\xi)\, da} \prod_{j=1}^{m}\left[1 + \frac{h(y_j|\xi)}{\epsilon}\right]$. However, one may often do better by exploiting specific model characteristics, as we shall illustrate in Section 4.4.6 below. If we couple the spatial birth-and-death process defined by $\bar{b}(\cdot)$ and $\underline{d}(\cdot, \cdot)$ to processes defined by $b(\cdot, \cdot)$ and $d(\cdot, \cdot)$ as in Kendall and Møller (2000), we obtain the following algorithm.

Algorithm 1 Let $V_{t,\xi}$, $t \leq 0$, $\xi \in \mathcal{X}$, be a family of independent, uniformly distributed random variables on $(0, 1)$. Initialise $T = 1$, and let $D(0)$ be a sample from $\pi(\cdot)$. Repeat

- extend $D(\cdot)$ backwards until time $-T$ by means of a spatial birth-and-death process with birth rate $\bar{b}(\cdot)$ and death rate $\underline{d}(\cdot, \cdot)$;

- generate a lower process $L_{-T}(\cdot)$ and an upper process $U_{-T}(\cdot)$ on $[-T, 0]$ as follows:

  - initialise $L_{-T}(-T) = \emptyset$, $U_{-T}(-T) = D(-T)$;
  - to each forward transition time $t \in (-T, 0]$ of $D(\cdot)$ correspond updates of the upper and lower processes;
  - in case of a death (i.e. a backwards birth), say $D(t) = D(t-) \setminus \{d\}$ where $D(t-)$ denotes the state just prior to time t, the point d is deleted from $L_{-T}(t-)$ and $U_{-T}(t-)$ as well;
  - in case of a birth, say $D(t) = D(t-) \cup \{\xi\}$, the point ξ is added to $U_{-T}(t-)$ only if

$$V_{t,\xi} \leq \max \left\{ \frac{b(\mathbf{x}, \xi)\, \underline{d}(\mathbf{x} \cup \{\xi\}, \xi)}{\overline{b}(\xi)\, d(\mathbf{x} \cup \{\xi\}, \xi)} : L_{-T}(t-) \subseteq \mathbf{x} \subseteq U_{-T}(t-) \right\}.$$

    similarly, ξ is added to $L_{-T}(t-)$ only if $V_{t,\xi}$ does not exceed the above expression with a minimum instead of a maximum;

- if $U_{-T}(0) = L_{-T}(0)$, return the common value $U_{-T}(0)$; otherwise set $T := 2T$;

  until the upper and lower processes have coalesced.

The next theorem gives conditions for algorithm 1 to output unbiased samples from the posterior distribution of cluster centres.

Theorem 1 Let $p_X(\cdot)$ be an anti-monotone, locally stable Markov point process density with respect to a unit rate Poisson process on a compact set $\mathcal{X} \subseteq \mathbb{R}^d$, and $h(\cdot|\cdot)$ a uniformly bounded displacement function of a Cox cluster process U observed in a bounded window A. Suppose the birth rates $b(\cdot, \cdot)$ and death rates $d(\cdot, \cdot)$ define a unique spatial birth-and-death process converging in distribution to the posterior density of cluster centres $p_{X|Y}(\cdot \mid \mathbf{y})$, and there exist upper and lower bounds (4.24)–(4.25) also defining a unique spatial birth-and-death process that converges in distribution to a probability density $\pi(\cdot)$ for which $\pi(\emptyset) > 0$ and detailed balance between births and deaths holds. Then the coupling from the past algorithm 1 almost surely terminates and outputs an unbiased sample from $p_{X|Y}(\cdot \mid \mathbf{y})$.

The proof is an adaptation to the inhomogeneous case of the proof in Kendall and Møller (2000).

Proof. First, note that by assumption the dominating process $D(\cdot)$ is in equilibrium, its distribution being defined by $\pi(\cdot)$. Clearly, for all $T > 0$,

$$\emptyset = L_{-T}(-T) \subseteq U_{-T}(-T) = D(-T)$$

and by construction the updates respect the inclusion order. Hence $L_{-T}(t) \subseteq U_{-T}(t)$ for all $t \in [-T, 0]$. Moreover, the processes funnel, i.e.

$$L_{-T}(t) \subseteq L_{-S}(t) \subseteq U_{-S}(t) \subseteq U_{-T}(t) \qquad (4.26)$$

whenever $-S \leq -T \leq t \leq 0$. The first inclusion can be verified by noting that $L_{-T}(-T) = \emptyset \subseteq L_{-S}(-T)$ and recalling that the transitions respect the inclusion order. Since $U_{-T}(-T) = D(-T) \supseteq U_{-S}(-T)$, the last inclusion in (4.26) follows by the same argument. If $L_{-T}(t_0) = U_{-T}(t_0)$ for some $t_0 \in [-T, 0]$, as the processes are coupled, $L_{-T}(t) = U_{-T}(t)$ for all $t \in [t_0, 0]$. Next, set $X_{-T}(-T) = \emptyset$ and define a process $X_{-T}(\cdot)$ on $[-T, 0]$ in analogy to the upper and lower processes, except that if $X_{-T}(t-) = \mathbf{x}$ the birth at time t of a point ξ is accepted if $V_{t,\xi} \leq \frac{b(\mathbf{x},\xi)\, d(\mathbf{x} \cup \{\xi\}, \xi)}{\bar{b}(\xi)\, d(\mathbf{x} \cup \{\xi\}, \xi)}$. In other words, $X_{-T}(\cdot)$ exhibits the dynamics of a spatial birth-and-death process with birth rate $\tilde{b}(\mathbf{x}, \xi) = b(\mathbf{x}, \xi)\frac{d(\mathbf{x} \cup \{\xi\}, \xi)}{d(\mathbf{x} \cup \{\xi\}, \xi)}$ and death rate $\tilde{d}(\mathbf{x} \cup \{\xi\}, \xi) = \underline{d}(\mathbf{x} \cup \{\xi\}, \xi)$. Thus, its detailed balance equations coincide with those for $b(\cdot, \cdot)$ and $d(\cdot, \cdot)$. Furthermore, $\tilde{b}(\cdot, \cdot) \leq b(\cdot, \cdot)$, hence explosion is prevented so that the process converges in distribution to its equilibrium distribution defined by $p_{X|Y}(\cdot \mid \mathbf{y})$. The inclusion properties derived above imply $L_{-T}(0) \subseteq X_{-T}(0) \subseteq U_{-T}(0)$, so that – provided the sampler terminates almost surely – with probability 1 the limit $\lim_{T \to \infty} X_{-T}(0)$ is well-defined. Since $D(\cdot)$ is in equilibrium, $X_{-T}(0)$ has the same distribution as if the X-process were run forward from time 0 (coupled to the dominating process as before) over a time period of length T, the limit distribution of which is $p_{X|Y}(\cdot \mid \mathbf{y})$. We conclude that the algorithm outputs an unbiased sample from the posterior distribution of parents. It remains to show that coalescence occurs almost surely. Recall that by assumption $\pi(\emptyset) > 0$. Set, for $n \in \mathbb{N}_0$, $E_n = \mathbf{1}\{D(-n) \neq \emptyset\}$. Now $(E_n)_n$ is an irreducible aperiodic Markov chain on $\{0, 1\}$ for which the equilibrium probability $\pi(\emptyset)$ of state 0 is strictly positive. Hence state 0 will be reached with probability 1, which implies that the dominating process $D(t)_{t \leq 0}$ will almost surely be empty for some t. But then (4.26) and the coupling imply that the algorithm terminates almost surely, and the proof is complete. $\square$

4.4.6 Example: Cox–Matérn Cluster Process

To describe a tailor-made coupling from the past algorithm, consider a Cox cluster process with intensity function given by (4.4)–(4.5) and prior density (4.17). For this model, the birth rate (4.20) satisfies

$$b(\mathbf{x}, \xi) \leq \beta \left[1 + \sum_{j=1}^{m} \frac{h(y_j | \xi)}{\epsilon} \right] = \bar{b}(\xi). \qquad (4.27)$$

In order to derive a lower bound for the death rate (4.21), note that

$$1 + \frac{h(y_j|\xi)}{\lambda(y_j \mid \mathbf{x})} \leq 1 + \frac{\mu}{\epsilon + \mu} \leq 2$$

if $y_j \in B(\xi, R_h) \cap U_{\mathbf{x}}$, where $U_{\mathbf{x}} = \cup_{x_i \in \mathbf{x}} B(x_i, R_h)$ denotes the union of balls centred at the points of $\mathbf{x}$. It follows that the death rate $d(\mathbf{x} \cup \{\xi\}, \xi)$ is bounded below by

$$\underline{d}(\mathbf{x} \cup \{\xi\}, \xi) = d(\mathbf{x} \cup \{\xi\}, \xi) \prod_{j:y_j \in B(\xi, R_h) \cap U_{\mathbf{x}}} \left(\frac{1}{2} + \frac{\mu}{2\,\lambda(y_j \mid \mathbf{x})} \right). \quad (4.28)$$

By the Preston theorem, the transition rates $\bar{b}(\xi)$ and $\underline{d}(\mathbf{x} \cup \{\xi\}, \xi)$ define

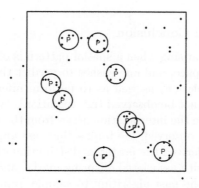

Figure 4.5 *Realisation of extrapolated redwood seedling pattern on $\mathcal{X} \oplus B(0, R_h)$ from observations in unit square (box) and interpolated parent pattern ('P') based on a Cox–Matérn cluster process with $R_h = 0.061$ and $(\hat{\epsilon}_{100}, \hat{\mu}_{100})$ obtained by coupling from the past. We used a hard core prior with $R = 0.03$ and $\beta = 1.0$.*

a unique spatial birth-and-death process, whose limit distribution is given by

$$\pi(\mathbf{x}) \propto \gamma^{-n(\mathbf{y} \cap U_{\mathbf{x}})} \prod_{i=1}^{n(\mathbf{x})} \beta(x_i), \quad (4.29)$$

a generalised area-interaction process (Widom and Rowlinson 1970, Baddeley and van Lieshout 1995, Kendall 1998, Häggström et al. 1999) with intensity function

$$\beta(\xi) = \beta \exp\left[-\int_A h(a|\xi)\, da \right] 2^{n(\mathbf{y} \cap B(\xi, R_h))}$$

and interaction parameter $\gamma = (\frac{1}{2} + \frac{\mu}{2\epsilon})^{-1}$. Regarding the implementation of algorithm 1, note that

$$\frac{b(\mathbf{x}, \xi)\,\underline{d}(\mathbf{x} \cup \{\xi\}, \xi)}{\overline{b}(\xi)\,d(\mathbf{x} \cup \{\xi\}, \xi)} = \mathbf{1}\{d(\xi, \mathbf{x}) > R\} \prod_{j:y_j \in B(\xi, R_h) \cap U_\mathbf{x}} \left(\frac{1}{2} + \frac{\mu}{2\lambda(y_j \mid \mathbf{x})}\right)$$

is decreasing in $\mathbf{x}$, so the sampler is anti-monotone, and the births in the upper and lower processes may be implemented by simply considering the current state of the other process at each transition; see Kendall (1998). We applied the above algorithm to the redwood seedlings data of Figure 4.1 for the Matérn parameter vector (ϵ, μ) equal to its Monte Carlo maximum likelihood estimate (cf. Section 4.4.4) and a hard core prior with $\beta = 1.0$ and $R = 0.03$ as before. A typical realisation from the posterior distribution of parents can be seen in Figure 4.5 as well as an extrapolation of the redwood pattern to the set $\mathcal{X} \oplus B(0, R_h)$.

4.5 Summary and Conclusion

We discussed issues arising when a spatial pattern is observed within some bounded region of space, and one wishes to predict the process outside of this region (extrapolation) as well as to perform inference on features of the pattern that cannot be observed (interpolation). We focused on spatial cluster analysis. Here the interpolation arises from the fact that the centres of clustering are not observed. We took a Bayesian approach with a repulsive Markov prior, derived the posterior distribution of the complete data, i.e. cluster centres with associated offspring marks, and proposed an adaptive coupling from the past algorithm to sample from this posterior. The approach was illustrated by means of the redwood data set (Ripley 1977).

Acknowledgements

We gratefully acknowledge the expert programming support of Adri Steenbeek, very helpful comments from Yih Chong Chin, Nick Fisher, Richard Gill, Ilya Molchanov and Elke Thönnes, and the influence of previous unpublished collaboration with Andrew Lawson and Henry Y.W. Cheng. The research for this chapter was carried out under CWI project PNA4.3 'Stochastic Geometry', ARC large grant A69941083 'Extrapolating and Interpolating Spatial Patterns', and NWO grant CIMS 613.003.100.

CHAPTER 5

Perfect Sampling for Point Process Cluster Modelling

I.W. McKeague   *M. Loizeaux*

5.1 Introduction

When disease incidence locations are observed in a region, there is often interest in studying whether there is clustering about landmarks representing possible centralized sources of the disease. In this article we study a Bayesian approach to the detection and estimation of such landmarks. Spatial point processes are used to specify both the observation process and the prior distribution of the landmarks. We develop a perfect sampling algorithm for the posterior distribution of landmarks under various conditions on the prior and likelihood. Bayesian cluster models of the type we consider were introduced by Baddeley and van Lieshout (1993), primarily for applications in computer vision. The dissertation of van Lieshout (1995) (see also Chapter 4) focused on the special case of the Neyman–Scott cluster process, in which the observations arise from a superposition of inhomogeneous Poisson processes associated with each landmark (Neyman and Scott 1958) and she applied it to the well-known redwood seedling data used by Strauss (1975). Hurn (1998) applied the Baddeley–van Lieshout approach to the study of changes in the size and shape of living cells. Lawson and Clark (1999b) survey the statistical literature on disease clustering models. Markov chain Monte Carlo (MCMC) techniques are indispensable for the application of point process models in statistics, see, for example, the survey of Møller (1999). The typical MCMC sampler obtains draws that are at best only approximately from the target distribution, and are often plagued by convergence problems, even when a long "burn-in" period is used. Moreover, if independent draws are required, then every draw must be produced by a separate chain. However, using an algorithm developed by Kendall and Møller (2000), it is possible to sample perfectly from the posterior distribution in the Bayesian cluster model. Perfect samplers originate in the seminal work of Propp and Wilson (1996), whose coupling from the past (CFTP) algorithm delivers an *exact* draw from the target distribution. The most important practical advantage of perfect samplers over traditional MCMC schemes is that the need to assess convergence of the sampler is eliminated. There are many examples of perfect samplers in the literature. Mira et al. (2001) apply perfect simulation to slice samplers for bounded target distributions; Casella et al. (1999) create perfect slice samplers for mixtures of exponential distributions and mixtures of normal distributions. For an example of perfect sampling of a conditioned Boolean model, see Cai and Kendall (1999). Häggström et al. (1999) obtain perfect samples from the area–interaction point process (attractive case) using auxiliary variables. Another version of perfect sampling called read-once CFTP, which runs the Markov chain forward in time and never restarts it at previous past times (thus avoiding the need to store random numbers for reuse), is given by Wilson (2000). For recent applications of perfect simulation in statistics, see Green and Murdoch (1999) and Møller and Nicholls (1999). The Kendall–Møller version of perfect simulation was developed for spatial point processes that are locally stable. We show that the posterior distribution in the Bayesian cluster model is locally stable on its support, provided the prior is locally stable and the likelihood satisfies some mild conditions. In particular, this shows that the posterior density is *proper* (has unit total mass). The Kendall–Møller algorithm is computationally *feasible*, however, only when the locally stable point process is either attractive (favoring clustered patterns) or repulsive (discouraging clustered patterns), or a product of attractive and repulsive components. We examine the feasibility of the sampler in two special cases: the Neyman–Scott process and the pure silhouette model (in which only the support of the observations is determined by the landmarks). Our approach is applied to data on leukemia counts in an eight county area of upstate New York during the years 1978–82. There is an extensive literature on the analysis of these data, the main goal being the detection of disease clusters; see, for example, Ghosh et al. (1999), Ahrens et al. (1999) and Denison and Holmes (2001). The study area includes 11 inactive hazardous waste sites, and it is natural in any study to attempt to determine if any of these sites can be seen as a contributor to the incidence of leukemia in the area. We find evidence for an elevated leukemia incidence in the neighborhood of one of the sites. The paper is organized as follows. Section 5.2 gives background material and describes the general model. In this section we also state the main result of the paper showing that the posterior is locally stable on its support. Section 5.4 examines several examples of the basic model, and addresses the question of whether perfect sampling is feasible in each case. Section 5.5 contains the disease clustering application, and a further application to the classic redwood seedlings data is given in Section 5.6.

5.2 Bayesian Cluster Model

5.2.1 Preliminaries

The point process notation used throughout the paper is standard, cf. Häggström et al. (1999), Baddeley and Møller (1989), or Møller (1999, 2001b). For in-depth accounts of the theory of point processes, see Daley and Vere-Jones (1988) and van Lieshout (2000). Let W be a compact subset of the plane representing the study region, and let $|\cdot|$ denote Lebesgue measure on W. A realization of a point process in W is a finite set of points $\boldsymbol{x} = \{x_1, x_2, \ldots, x_{n(\boldsymbol{x})}\} \subset W$, where $n(\boldsymbol{x})$ is the number of points in $\boldsymbol{x}$. If $n(\boldsymbol{x}) = 0$, write $\boldsymbol{x} = \emptyset$ for the empty configuration. Let Ω denote the exponential space (Carter and Prenter 1972) of all such finite point configurations in W, and furnish it with the σ-field $\mathcal{F}$ generated by sets of the form $\{\boldsymbol{x} : n(\boldsymbol{x} \cap B) = k\}$, where $B \in \mathcal{B}$, the Borel σ-field on W, and $k = 0, 1, 2, \ldots$. Let $(W, \mathcal{B}, \lambda)$ be a measure space with $\lambda(W) < \infty$. The Poisson process Z on W with intensity measure λ is an Ω-valued random variable such that the number of points $n(Z)$ follows the Poisson distribution with mean $\lambda(W)$, and conditional on $n(Z) = m$, the m points are independent and identically distributed with distribution $\lambda(\cdot)/\lambda(W)$. If λ is Lebesgue measure $|\cdot|$, then we say Z is the unit rate Poisson process on W. All point processes considered here will have distributions that are absolutely continuous with respect to the distribution π of the unit rate Poisson process, and thus specified by a density with respect to π. Two important point processes are the Strauss (1975) process and the area-interaction process of Baddeley and van Lieshout (1995). The Strauss process has unnormalized density $f(\boldsymbol{x}) = \beta^{n(\boldsymbol{x})}\gamma^{t(\boldsymbol{x})}$, where $\beta > 0$, $0 < \gamma \leq 1$ and $t(\boldsymbol{x})$ is the number of unordered pairs of points in $\boldsymbol{x}$ which are within a specified distance r of each other. The Strauss process only models pairwise interaction which is repulsive, i.e., the points in the process tend to separate from one another. For the area-interaction process, the unnormalized density is given by $f(\boldsymbol{x}) = \beta^{n(\boldsymbol{x})}\gamma^{-\nu(U(\boldsymbol{x}))}$, where $\beta > 0$, $\gamma > 0$, ν is a totally finite, regular Borel measure on W, and $U(\boldsymbol{x})$ is the intersection of W with the union of balls of fixed radius r centered at points in $\boldsymbol{x}$. We can also write $U(\boldsymbol{x}) = (\boldsymbol{x} \oplus G) \cap W$, where $\oplus$ denotes Minkowski addition, and the grain G is the ball of radius r centered at the origin (more generally, G can be any compact subset of $\mathbb{R}^2$). Area-interaction processes can be used to model both clustered (attractive) and ordered (repulsive) patterns, and allow interactions of infinite order. The parameter γ controls the type of interaction; $\gamma > 1$ produces clustered patterns, $0 < \gamma < 1$ produces ordered patterns. An unnormalized density f (or corresponding process X) is said to be locally stable if there is a constant $K > 0$ such that

$$f(\boldsymbol{x} \cup \{\xi\}) \leq Kf(\boldsymbol{x})$$

for all $\boldsymbol{x} \in \Omega$, $\xi \in W \backslash \boldsymbol{x}$. As noted by Kendall and Møller (2000), local stability is equivalent to an upper bound on the Papangelou conditional intensity

$$q(\boldsymbol{x}, \xi) \equiv \frac{f(\boldsymbol{x} \cup \{\xi\})}{f(\boldsymbol{x})}, \ \boldsymbol{x} \in \Omega, \ \xi \in W \backslash \boldsymbol{x},$$

(with $c/0 = 0$ for $c \geq 0$) and the hereditary property, which states that if $\boldsymbol{x} \subset \boldsymbol{x}'$ then $f(\boldsymbol{x}') > 0$ implies $f(\boldsymbol{x}) > 0$. In addition, local stability implies $f(\emptyset) > 0$, since repeated use of the condition gives $f(\boldsymbol{x}) \leq K^{n(\boldsymbol{x})} f(\emptyset)$. Most point processes that have been suggested for modeling spatial point patterns are locally stable, including the Strauss and the area-interaction processes.

5.2.2 Model Specification

The observed point configuration which arises from the landmarks $\boldsymbol{x}$ will be denoted $\mathbf{y} = \{y_1, y_2, \ldots, y_{n(\mathbf{y})}\} \subset W$, and assumed to be non-empty. The prior and observation models are specified by point processes on W. The prior distribution of landmarks corresponds to a point process X having density $p_X(\mathbf{x})$ with respect to π. The landmarks produce daughters $\mathbf{y}$, a certain proportion of which will be required to fall in a silhouette region $S(\boldsymbol{x}) \subset W$ given by the union of discs

$$S(\boldsymbol{x}) = \cup_{\xi \in \boldsymbol{x}} D(\xi, r_{\text{sil}}) = (\boldsymbol{x} \oplus G_{\text{sil}}) \cap W,$$

where $D(\xi, r_{\text{sil}}) = \{x \in W : ||x - \xi|| \leq r_{\text{sil}}\}$, $|| \cdot ||$ is Euclidean distance, and G_{sil} is the disk of radius r_{sil} centered at the origin. The random closed set $S(X)$ is known as a Boolean model, see, for example, Chapter 3 of Stoyan et al. (1995). With appropriate measurability conditions, our approach allows more general types of silhouettes $S(\boldsymbol{x})$, but we have restricted attention to Boolean silhouettes for simplicity. We have used different notation for $S(\boldsymbol{x})$ and $U(\boldsymbol{x})$, even though they are silhouettes of the same type, to avoid confusion later on. The observation process is defined in terms of an unnormalized density $f(\cdot | \boldsymbol{x})$ conditioned on the event $\{\mathbf{y} : s(\mathbf{y} | \boldsymbol{x}) \geq c\}$, where $0 \leq c \leq 1$ and $s(\mathbf{y} | \boldsymbol{x}) = n(\mathbf{y} \cap S(\boldsymbol{x}))/n(\mathbf{y})$ is the proportion of points in $\mathbf{y}$ within the silhouette region (with $0/0 = 0$). Thus, for a given set of landmarks $\boldsymbol{x}$, the density of the observed point process Y with respect to π is

$$p_{Y|X=\boldsymbol{x}}(\mathbf{y}) = \alpha_Y(\boldsymbol{x}) f(\mathbf{y} | \boldsymbol{x}) 1\{s(\mathbf{y} | \boldsymbol{x}) \geq c\},$$

where

$$\alpha_Y(\boldsymbol{x}) = \left(\int_{\{\boldsymbol{v} : s(\boldsymbol{v} | \boldsymbol{x}) \geq c\}} f(\boldsymbol{v} | \boldsymbol{x}) \, \pi(d\boldsymbol{v}) \right)^{-1}$$

is the normalizing constant. We assume that $f(\mathbf{y} | \boldsymbol{x})$ is jointly measurable in $\boldsymbol{x}$ and $\mathbf{y}$. The silhouette portion of the observation process could be absorbed into $f(\mathbf{y} | \boldsymbol{x})$, as in the formulation of Baddeley and van Lieshout

(1993), but we find it useful to isolate $1\{s(\mathbf{y}|\boldsymbol{x}) \geq c\}$ in this fashion in order to provide conditions under which perfect sampling is feasible. Our formulation of the model also provides a flexible class of Boolean models which are of independent interest. When $f(\mathbf{y}|\boldsymbol{x}) = f(\mathbf{y})$ and $c = 1$ we say the likelihood follows a *pure silhouette model*; this is similar to the blur-free model of Baddeley and van Lieshout (1993). From Bayes formula, the posterior density of X with respect to π is

$$p_{X|Y=\mathbf{y}}(\boldsymbol{x}) \propto f_{X|Y=\mathbf{y}}(\boldsymbol{x})1\{s(\mathbf{y}|\boldsymbol{x}) \geq c\}, \tag{5.1}$$

where the unnormalized density

$$f_{X|Y=\mathbf{y}}(\boldsymbol{x}) = \alpha_Y(\boldsymbol{x})f(\mathbf{y}|\boldsymbol{x})p_X(\boldsymbol{x}) \tag{5.2}$$

is called the unrestricted posterior. Theorem 5.2.1 which follows provides sufficient conditions for the unrestricted posterior to be locally stable. We note in passing that the posterior itself is not hereditary (thus not locally stable) if $c > 0$, because $S(\emptyset) = \emptyset$ and $\mathbf{y} \neq \emptyset$. We assume local stability of the prior $p_X(\cdot)$ and of the likelihood $f(\mathbf{y}|\cdot)$ (for each fixed $\mathbf{y}$). In addition, we assume that the *family* of functions $\{f(\mathbf{y}|\cdot) : \mathbf{y} \in \Omega\}$ satisfies the following *local growth condition*: there exists a constant $L > 0$ such that

$$f(\mathbf{y}|\boldsymbol{x} \cup \{\xi\}) \geq Lf(\mathbf{y}|\boldsymbol{x}) \tag{5.3}$$

for all $\boldsymbol{x} \in \Omega$, $\xi \in W \backslash \boldsymbol{x}$. We call this a 'local growth' condition because it implies that the Papangelou conditional intensity for $f(\mathbf{y}|\cdot)$ is bounded away from zero. As it turns out, $L = 1$ in the examples to follow. Prior to stating our main theorem, for easy reference we state the following lemma giving the density of an inhomogeneous Poisson process with respect to the unit rate Poisson process. The proof can be found, for example, on page 499 of Daley and Vere-Jones (1988).

Lemma 5.2.1 Let π be the distribution of the unit rate Poisson process on the compact set $W \subset \mathbb{R}^2$, and let π^β be the distribution of the inhomogeneous Poisson process on W with intensity function $\beta(t)$. Then $\pi^\beta \ll \pi$ and

$$\frac{d\pi^\beta}{d\pi}(\boldsymbol{x}) = \exp\left(\int_W (1 - \beta(t))\, dt\right) \prod_{j=1}^{n(\boldsymbol{x})} \beta(x_j).$$

Theorem 5.2.1 Suppose $p_X(\cdot)$ and $f(\mathbf{y}|\cdot)$ (for each $\mathbf{y}$) are locally stable, and $\{f(\mathbf{y}|\cdot): \mathbf{y} \in \Omega\}$ satisfies the local growth condition (5.3). Then the unrestricted posterior (5.2) is locally stable.

Proof: We first show that density (5.1) is measurable. Since $p_X(\cdot)$ and $f(\mathbf{y}|\cdot)$ are measurable, we need only establish the measurability of $\alpha_Y(\cdot)$

and $\{x\colon s(y|x) \geq c\}$. First consider $\{x\colon s(y|x) \geq c\}$ in the case $c = 1$. With $y = \{y_1, y_2, \ldots, y_m\}$,

$$\{x\colon s(y|x) \geq 1\} = \{x\colon y \subset S(x)\} = \bigcap_{j=1}^{m}\{x\colon n(x \cap D(y_j, r)) \geq 1\}$$

$$= \bigcap_{j=1}^{m}\bigcup_{k=1}^{\infty}\{x\colon n(x \cap D(y_j, r)) = k\},$$

which belongs to $\mathcal{F}$. The general case reduces to the special case of $c = 1$ because $\{x\colon s(y|x) \geq c\}$ is the finite union of sets of the form $\{x\colon z \subset S(x)\}$ with $z \subset y$ and $n(z) \geq cn(y)$. To show that $\alpha_Y(\cdot)$ is measurable, the π-λ theorem allows us to reduce to the case $f(v|x) = 1_C(v)1_D(x)$, where C and D are generating sets in $\mathcal{F}$. Thus we only need show the measurability of

$$x \longmapsto \pi\{v\colon n(v \cap B) = k \text{ and } n(v \cap S(x)) \geq cn(v)\},$$

where $B \in \mathcal{B}$ and $k \geq 0$. Since π is Poisson, the above function only depends on x through a continuous function of the areas of the disjoint regions $B^* \cap S(x)^*$, where $A^* = A$ or A^c. These areas are measurable functions of x by Lemma 2.1 of Baddeley and van Lieshout (1995). To establish local stability of the unrestricted posterior it is enough to show that $\alpha_Y(\cdot)$ is locally stable; $p_X(\cdot)$ and $f(y|\cdot)$ were assumed to be locally stable, and finite products of locally stable densities are locally stable. Given $\xi \in W\backslash x$,

$$\begin{aligned}
\alpha_Y(x \cup \{\xi\})^{-1} &= \int_{\{v\colon s(v|x\cup\{\xi\})\geq c\}} f(v|x \cup \{\xi\})\,\pi(dv) \\
&\geq L\int_{\{v\colon s(v|x\cup\{\xi\})\geq c\}} f(v|x)\,\pi(dv) \\
&\geq L\int_{\{v\colon s(v|x)\geq c\}} f(v|x)\,\pi(dv) \\
&= L\alpha_Y(x)^{-1},
\end{aligned}$$

where we have used the local growth condition (5.3) and the fact that $s(v|x) \leq s(v|x \cup \{\xi\})$. Thus $\alpha_Y(\cdot)$ is locally stable. $\square$

It is convenient to introduce some notation for the bounds implied by the local stability and local growth conditions. Let K_{prior} be an upper bound on the Papangelou conditional intensity for $p_X(x)$, and let $K_{\text{lik}}(y)$ be an upper bound on the Papangelou conditional intensity for $f(y|\cdot)$. Under the assumptions given in Theorem 5.2.1, since finite products of locally stable densities are locally stable, we find that

$$K_{\text{post}} = \frac{1}{L}K_{\text{lik}}(y)K_{\text{prior}} \qquad (5.4)$$

is an upper bound on the Papangelou conditional intensity for the unrestricted posterior density (5.2).

5.3 Sampling from the Posterior

Clearly, once we develop a sampler for the unrestricted posterior (5.2), simple rejection sampling can be used to obtain samples from the posterior (5.1) itself. This is an efficient approach provided draws from the unrestricted posterior are not rejected too frequently. As previously stated, MCMC methods are needed to sample from (5.2), and from point processes in general. The Hastings–Metropolis–Green sampler of Geyer and Møller (1994) is applicable and easy to implement. Under the conditions of Theorem 5.2.1, the unrestricted posterior is locally stable, so the Geyer–Møller algorithm is geometrically ergodic; see Geyer (1999). As explained in the Introduction, however, it would be preferable to utilize the algorithm of Kendall and Møller (2000) if possible. This algorithm is not feasible in every situation, but when feasible has the advantage of producing perfect samples. Kendall (1998) introduced the algorithm initially for area-interaction processes. Kendall and Thönnes (1999) give a review of the method from the perspective of image processing. An alternative perfect sampler for area-interaction processes has been suggested by Häggström et al. (1999). The Kendall–Møller algorithm uses the method of dominated coupling from the past (CFTP) to produce perfect samples from a locally stable point process. The sampler is driven by a spatial birth-and-death process $\tilde{X}(t)$ having equilibrium distribution which agrees with the distribution of the target point process. Dominated CFTP is sometimes referred to as "horizontal CFTP", as opposed to the "vertical CFTP" originally introduced by Propp and Wilson (1996). Vertical CFTP requires minimal and maximal elements with respect to some partial ordering of the state space; it is not applicable to point processes because a maximal element does not exist in Ω under set inclusion. Horizontal CFTP, however, only requires a minimal element $\emptyset$ which functions as an ergodic atom for a dominating process $D(t)$. Once the dominating process $D(t)$ hits the minimal element, subsequent evolution of $\tilde{X}(t)$ does not depend on the initial state, so $\tilde{X}(0)$ represents the endpoint of a path from the infinite past (coalescence), and thus a perfect sample from the target distribution. A further aspect of horizontal CFTP is the use of lower and upper processes that bound $\tilde{X}(t)$ more efficiently than $\emptyset$ and $D(t)$, and thus provide coalescence more rapidly. These various processes are defined below. The spatial birth-and-death process $\tilde{X} = \{\tilde{X}(t) : t \in \mathbb{R}\}$ has birth rate $b(\boldsymbol{x}, \xi) = q(\boldsymbol{x}, \xi)$ given by the Papangelou conditional intensity corresponding to $f_{X|Y=\mathbf{y}}(\boldsymbol{x})$, and death rate $d(\boldsymbol{x}, \xi) = 1$. It is clear that the target density $f_{X|Y=\mathbf{y}}(\boldsymbol{x})$ and $\tilde{X}$

are in detailed balance, i.e.

$$f_{X|Y=\mathbf{y}}(\boldsymbol{x})b(\boldsymbol{x},\xi) = f_{X|Y=\mathbf{y}}(\boldsymbol{x} \cup \xi)d(\boldsymbol{x},\xi) > 0,$$

so $\tilde{X}(t)$ converges in distribution to X as $t \to \infty$, where X has density (5.2); see Kendall and Møller (2000). The dominating process $D(t)$ has birth rate K_{post} and death rate 1, so its equilibrium distribution is homogeneous Poisson with intensity K_{post}. Suppose that $\tilde{X}(t^-) = \boldsymbol{x} \subset \mathbf{y} = D(t^-)$ are the states of $\tilde{X}$ and D just before time t. If at time t a point ξ is added to D, then this same point is added to $\tilde{X}$ with probability $q(\boldsymbol{x},\xi)/K_{\text{post}}$, that is, if $M(t) \le q(\boldsymbol{x},\xi)/K_{\text{post}}$ where $M(t)$ is an iid uniform$(0,1)$ mark process. If at time t a point is deleted from D, then this same point is deleted from $\tilde{X}$. For every $n > 0$, the upper process U_{-n} is initialized at time $-n$ with $\hat{\boldsymbol{x}} = D(-n)$. The initial state for L_{-n} is the empty configuration. Using the mark $M(t)$, a new point ξ is added to U_{-n} at time $-t$, $t < n$, with probability

$$\max\left\{\frac{q(\boldsymbol{x},\xi)}{K_{post}} : L_{-n}^{-t^-}(\emptyset) \subseteq \boldsymbol{x} \subseteq U_{-n}^{-t^-}(\hat{\boldsymbol{x}})\right\},$$

and the same point ξ is added to L_{-n} with probability

$$\min\left\{\frac{q(\boldsymbol{x},\xi)}{K_{post}} : L_{-n}^{-t^-}(\emptyset) \subseteq \boldsymbol{x} \subseteq U_{-n}^{-t^-}(\hat{\boldsymbol{x}})\right\},$$

where $L_{-n}^{-t^-}(\emptyset)$ and $U_{-n}^{-t^-}(\hat{\boldsymbol{x}})$ are the states of $L_{-n}(\emptyset)$ and $U_{-n}(\hat{\boldsymbol{x}})$ just before time $-t$. The construction described then gives

$$L_{-n}^{-t}(\emptyset) \subseteq \tilde{X}(-t) \subseteq U_{-n}^{-t}(\hat{\boldsymbol{x}}) \subseteq D(-t),$$

and the funnelling property

$$L_{-n}^{-t} \le L_{-m}^{-t} \le U_{-m}^{-t} \le U_{-n}^{-t}, \ m \ge n \ge t.$$

Since X is locally stable by Theorem 5.2.1, the empty configuration is an ergodic atom for $D(t)$. Thus it is simply a matter of finding n large enough so that $L_{-n}^0(\emptyset) = U_{-n}^0(\hat{\boldsymbol{x}})$, in which case $L_{-n}^0(\emptyset)$ is a perfect draw from the posterior distribution. Note the inherent difficulty in calculating the probability of adding a point to U at time $-t$ (and similarly to L): in the general case it is necessary to evaluate $q(\boldsymbol{x},\xi)/K_{post}$ for every $\boldsymbol{x}$ such that $L_{-n}^{-t^-}(\emptyset) \subseteq \boldsymbol{x} \subseteq U_{-n}^{-t^-}(\hat{\boldsymbol{x}})$. These probabilities are easy to calculate, however, if the locally stable point process X is either attractive:

$$q(\boldsymbol{x},\xi) \le q(\boldsymbol{x}',\xi) \text{ whenever } \xi \notin \boldsymbol{x}' \text{ and } \boldsymbol{x} \subset \boldsymbol{x}',$$

or repulsive:

$$q(\boldsymbol{x},\xi) \ge q(\boldsymbol{x}',\xi) \text{ whenever } \xi \notin \boldsymbol{x}' \text{ and } \boldsymbol{x} \subset \boldsymbol{x}'.$$

5.4 Specialized Examples

In this section we examine various choices of the prior and likelihood which satisfy Theorem 5.2.1. In addition, we describe conditions under which the posterior is either attractive or repulsive, so that perfect simulation is feasible. A small dataset is used to illustrate the viability of perfect simulation in each case. In Section 5.4.1 the Neyman–Scott model is described in detail (see also Chapter 3). Section 5.4.2 provides several cases of the pure silhouette model.

5.4.1 Neyman–Scott Model

In this section we consider the Neyman–Scott model in which the observation process Y is the superposition of $n(\boldsymbol{x})$ independent inhomogeneous Poisson processes Z_{x_i} and a background Poisson noise process of intensity $\epsilon > 0$. The intensity $h(\cdot|x_i)$ of Z_{x_i} is specified parametrically, and the prior $p_X(\boldsymbol{x})$ is assumed to be locally stable. This fits into our general framework by taking $c = 0$ and $f(\mathbf{y}|\boldsymbol{x}) = \prod_{j=1}^{n(\mathbf{y})} \mu(y_j|\boldsymbol{x})$, where

$$\mu(t|\boldsymbol{x}) = \epsilon + \sum_{i=1}^{n(\boldsymbol{x})} h(t|x_i)$$

is the conditional intensity at t of Y given $\boldsymbol{x}$. We now check the relevant conditions of Theorem 5.2.1, assuming that $h(t|\cdot)$ is bounded for each $t \in W$. To show that $f(\mathbf{y}|\cdot)$ is locally stable, note that for $\xi \in W\backslash\boldsymbol{x}$

$$
\begin{aligned}
f(\mathbf{y}|\boldsymbol{x} \cup \{\xi\}) &= \prod_{j=1}^{n(\mathbf{y})} \left(\epsilon + \sum_{i=1}^{n(\boldsymbol{x})} h(y_j|x_i) + h(y_j|\xi) \right) \\
&\leq \prod_{j=1}^{n(\mathbf{y})} \left(\epsilon \left(1 + \frac{h(y_j|\xi)}{\epsilon} \right) + \sum_{i=1}^{n(\boldsymbol{x})} h(y_j|x_i) \right) \\
&\leq K_{\text{lik}}(\mathbf{y})f(\mathbf{y}|\boldsymbol{x}),
\end{aligned}
$$

where

$$K_{\text{lik}}(\mathbf{y}) = \sup_{\xi \in W} \prod_{j=1}^{n(\mathbf{y})} \left(1 + \frac{h(y_j|\xi)}{\epsilon} \right) < \infty.$$

To check the local growth condition, note that for $\xi \in W\backslash\boldsymbol{x}$

$$f(\mathbf{y}|\boldsymbol{x} \cup \{\xi\}) = \prod_{j=1}^{n(\mathbf{y})} (\mu(y_j|\boldsymbol{x}) + h(y_j|\xi)) \geq f(\mathbf{y}|\boldsymbol{x}),$$

uniformly in $\mathbf{y}$, and we can use $L = 1$. Thus the conditions of Theorem 5.2.1 are satisfied, so the posterior is locally stable. As previously noted, the

Kendall–Møller algorithm is feasible if the posterior is either attractive or repulsive, or factors into attractive/repulsive components. The Papangelou conditional intensity corresponding to $f(\mathbf{y}|\cdot)$ is

$$\frac{f(\mathbf{y}|\boldsymbol{x} \cup \{\xi\})}{f(\mathbf{y}|\boldsymbol{x})} = \prod_{j=1}^{n(\mathbf{y})} \left(1 + \frac{h(y_j|\xi)}{\epsilon + \sum_{i=1}^{n(\boldsymbol{x})} h(y_j|x_i)} \right),$$

for $\xi \in W \backslash \boldsymbol{x}$, which is clearly decreasing in $\boldsymbol{x}$, thus repulsive. By Lemma 5.2.1 (with intensity $\beta(t) = \mu(t|\boldsymbol{x})$), we have

$$\alpha_Y(\boldsymbol{x}) = \exp \left\{ \int_W (1 - \mu(t|\boldsymbol{x}))\, dt \right\}.$$

Therefore, the Papangelou conditional intensity corresponding to $\alpha_Y(\cdot)$ is

$$\frac{\alpha_Y(\boldsymbol{x} \cup \{\xi\})}{\alpha_Y(\boldsymbol{x})} = \exp \left\{ -\int_W h(t|\xi)\, dt \right\},$$

for $\xi \in W \backslash \boldsymbol{x}$, which does not depend on $\boldsymbol{x}$. We conclude that the posterior is repulsive whenever the prior is repulsive, and sampling from the posterior is possible if the prior is either repulsive or attractive. Figure 5.1 shows an artificial dataset in the unit square consisting of two clusters of three observations each and one isolated observation. Figure 5.2 gives an illustration of perfect sampling for the Neyman–Scott model based on these data. We used a Strauss prior that produces empty configurations of landmarks approximately 50% of the time. For the likelihood we used the Thomas

Figure 5.1 *Artificial dataset used to illustrate the examples in Section 1.4.*

intensity model (radially symmetric Gaussian)

$$h(t|x) = \frac{\kappa}{2\pi\sigma^2} e^{-||t-x||^2/2\sigma^2}, \tag{5.5}$$

where $\kappa, \sigma > 0$. Note that the contribution of a particular landmark x to the intensity rate at x is exactly $\kappa^* = \kappa/(2\pi\sigma^2)$, the maximum value of $h(t|x)$. We will refer to κ^* as the (Thomas) landmark weight, and find it convenient to express κ^* as a multiple of the background intensity rate ϵ. Large values

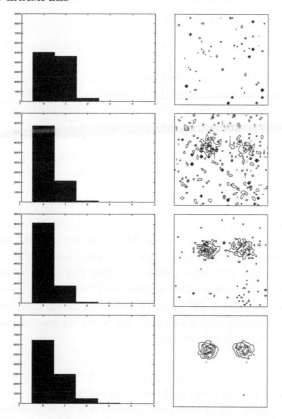

Figure 5.2 *Features of the prior and posterior distributions using the Neyman–Scott model with a Thomas intensity, $\epsilon = 0.1$ and $\sigma = 0.07$, and a Strauss prior, $\beta_X = 0.1$, $r = 0.1$, and $\gamma_X = 0.1$. The dataset consists of seven points in the unit square. Left: histogram of the number of landmarks. Right: contour plot of the intensity. Top: the Strauss prior. $\epsilon = 0.1$ and $\sigma = 0.07$. Row 2: the posterior with $\kappa^* = \epsilon$. Row 3: the posterior with $\kappa^* = 2\epsilon$. Row 4: the posterior with $\kappa^* = 5\epsilon$. In each case, 10,000 exact samples were drawn using the Kendall–Møller algorithm.*

of κ^* increase the influence of the landmarks, while small values increase the influence of the background noise. The Kendall–Møller algorithm was used to sample from the posterior; $K_{\text{prior}} = \beta_X$. The algorithm provided 10,000 samples on a Silicon Graphics workstation in less than 1 minute for the prior and about 3 minutes for the largest value of κ^*.

5.4.2 Pure Silhouette Models

In this section we look at some special cases of the pure silhouette model:
$f(\mathbf{y}|\boldsymbol{x}) = f(\mathbf{y})$ and $c = 1$. In this case, $\{s(\mathbf{y}|\boldsymbol{x}) \geq c\} = \{\mathbf{y} \subset S(\boldsymbol{x})\}$,
and the landmarks are now unidentifiable in a frequentist sense because
the likelihood only depends on the landmarks through their silhouette: for
$\xi \in W\backslash\boldsymbol{x}$ such that $S(\{\xi\}) \subset S(\boldsymbol{x})$ (that is, $S(\boldsymbol{x} \cup \{\xi\}) = S(\boldsymbol{x})$), we have
$p_{Y|X=\boldsymbol{x}}(\cdot) = p_{Y|X=\boldsymbol{x}\cup\{\xi\}}(\cdot)$. The conditions of Theorem 5.2.1 are satisfied
provided the prior is locally stable, and then $K_{\text{post}} = K_{\text{prior}}$. The posterior
now only depends on the data through the restriction $\mathbf{y} \subset S(\boldsymbol{x})$ because
$f(\mathbf{y})$ is absorbed into the normalizing constant:

$$p_{X|Y=\mathbf{y}}(\boldsymbol{x}) \propto \alpha_Y(\boldsymbol{x})p_X(\boldsymbol{x})1\{\mathbf{y} \subset S(\boldsymbol{x})\}.$$

Thus it is possible to generate draws from the unrestricted posterior before
seeing the data. The computationally expensive MCMC simulations can
be carried out ahead of time; once the data become available, we just
select those draws with silhouettes covering the data. Suppose now that
the observation process is inhomogeneous Poisson with intensity $\beta_Y(t)$ on
the silhouette, so $f(\mathbf{y}) = \prod_{j=1}^{n(\mathbf{y})} \beta_Y(y_j)$. If the prior is homogeneous Poisson
then the unrestricted posterior turns out to be an area-interaction process,
as stated in the following theorem.

Theorem 5.4.1 In the pure silhouette model, if the prior is homogeneous
Poisson with parameter β_X, and the observation process is Poisson with intensity $\beta_Y(\cdot)$, then the unrestricted posterior is an area-interaction process
with $\nu(\cdot) = \int_{\cdot} \beta_Y(t)\,dt$ and parameters $\beta = \beta_X$, $\gamma = e$:

$$p_{X|Y=\mathbf{y}}(\boldsymbol{x}) \propto \beta_X^{n(\boldsymbol{x})}e^{-\nu(S(\boldsymbol{x}))}1\{\mathbf{y} \subset S(\boldsymbol{x})\}.$$

Proof: Note that

$$\alpha_Y(\boldsymbol{x}) = \left[\int_{\{\boldsymbol{v}:\boldsymbol{v}\subset S(\boldsymbol{x})\}} \prod_{j=1}^{n(\boldsymbol{v})} \beta_Y(v_j)\,\pi(d\boldsymbol{v})\right]^{-1}$$

$$= \left[\int_W \prod_{j=1}^{n(\boldsymbol{v})} \beta_Y(v_j)1\{v_j \in S(\boldsymbol{x})\}\,\pi(d\boldsymbol{v})\right]^{-1}.$$

Then, applying Lemma 5.2.1 to the intensity $\beta(t) = \beta_Y(t)1\{t \in S(\boldsymbol{x})\}$,

$$\alpha_Y(\boldsymbol{x}) = \exp\left\{|W| - \int_{S(\boldsymbol{x})} \beta_Y(t)\,dt\right\}.$$

Since $\gamma_X = 1$ and $e^{|W|}$ can be absorbed into the normalizing constant, the result follows by substitution in (5.1). $\square$

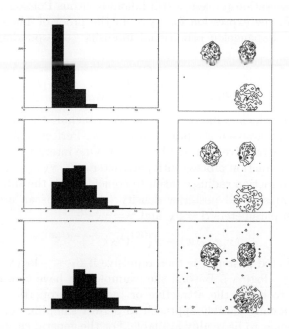

Figure 5.3 *Features of the posterior using the pure silhouette model, $r_{\text{sil}} = 0.15$, and a Poisson prior. The observation model is homogeneous Poisson, with $\beta_Y = 3$. Top row: $\beta_X = 0.693$, so that the prior probability of the empty configuration is 0.5. Middle row: $\beta_X = 2$. Bottom row: $\beta_X = 3$. In each case 500 exact samples were drawn using the Kendall–Møller algorithm.*

Thus the unrestricted posterior is attractive ($\gamma > 1$) in this case. For a homogeneous Poisson likelihood with intensity β_Y, the above result reduces to

$$p_{X|Y=\mathbf{y}}(\boldsymbol{x}) \propto \beta_X^{n(\boldsymbol{x})}(e^{\beta_Y})^{-|S(\boldsymbol{x})|}1\{\mathbf{y} \subset S(\boldsymbol{x})\}. \tag{5.6}$$

To implement perfect sampling for (5.6), use $K_{\text{post}} = K_{\text{prior}} = \beta_X$. Figure 5.3 illustrates this example with the same data as Figure 5.2. For this figure we varied the prior intensity β_X, while keeping all other hyperparameters constant. In the top row the prior probability of the empty configuration is 0.5 ($\beta_X = 0.693$). The Kendall–Møller algorithm along with rejection sampling provided 500 exact samples in about 72 hours on a Silicon Graphics workstation. The sampling in this case took much longer than for the Neyman–Scott model (even for a reduced number of samples) due to an acceptance rate in the rejection sampling step of approximately $1/44,000$; this was caused by the difficulty of covering the data by the silhouette of the landmarks. The results are markedly different from the Neyman–Scott model (Figure 5.2): three distinct circular clusters are readily apparent. Now suppose that the prior is an arbitrary locally stable point process (and the observation process is still inhomogeneous Poisson with intensity $\beta_Y(t)$). From the expression for $\alpha_Y(\cdot)$ in the proof of Theorem 5.4.1, we see that the Papangelou conditional intensity corresponding to $\alpha_Y(\cdot)$ is attractive:

$$\frac{\alpha_Y(\boldsymbol{x} \cup \{\xi\})}{\alpha_Y(\boldsymbol{x})} = \exp\left(-\int_{S(\{\xi\})\backslash S(\boldsymbol{x})} \beta_Y(t)\, dt\right)$$

for $\xi \in W \backslash \boldsymbol{x}$, which is increasing in $\boldsymbol{x}$. We conclude that the unrestricted posterior is attractive if the prior is attractive. Perfect sampling is feasible for either an attractive or a repulsive prior. One interesting choice of prior is the area-interaction process with parameters $\beta_X, \gamma_X, \nu = $ Lebesgue measure, and interaction radius r taken to coincide with the silhouette radius. Then the unrestricted posterior density (cf. (5.6)) is an area-interaction process with parameters $\beta = \beta_X$ and $\gamma = \gamma_X e^{\beta_Y}$:

$$f_{X|Y=\mathbf{y}}(\boldsymbol{x}) = \beta_X^{n(\boldsymbol{x})}(\gamma_X e^{\beta_Y})^{-|U(\boldsymbol{x})|}. \tag{5.7}$$

The unrestricted posterior is then repulsive if $\beta_Y \leq -\log(\gamma_X)$ and attractive if $\beta_Y \geq -\log(\gamma_X)$. Beyond the examples we have considered, it may be difficult to check the attractive or repulsive properties; in that case, a computationally feasible approach to perfect sampling for the posterior does not appear to be readily available. For the general cluster model with $0 < c < 1$, the limits of integration for $\alpha_Y(\boldsymbol{x})$ are difficult to handle. If $c = 0$ this integral is over the entire window W. If $c = 1$ this integral is over the silhouette $S(\boldsymbol{x})$. But for the general case no such simplification seems available, and the resulting Papangelou conditional intensity is intractable. A fully Bayesian treatment in which the various hyperparameters are assigned priors may also be difficult to implement using perfect sampling.

5.5 Leukemia Incidence in Upstate New York

The study region is comprised of 790 census tracts in an eight county region of upstate New York; it includes the cities of Syracuse in the north and Binghamton in the south. The 1980 U.S. census reported a total of 1,057,673 residents. Leukemia incidence was recorded by the New York Department of Health during the years 1978–82, see Waller et al. (1992,1994). The data are available from the Statlib archive (lib.stat.cmu.edu). In some instances a case could not be associated with a unique census tract, resulting in many fractional counts. Our approach does not accommodate this type of data, so we follow Ghosh et al. (1999) and group the 790 census tracts into 281 blocks in order to identify most of the cases with a specific block. Less than 10% of all cases could not be identified with a specific block, and such cases are excluded from our analysis. The locations of the centroids of the

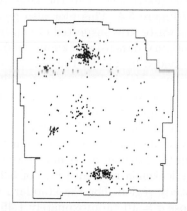 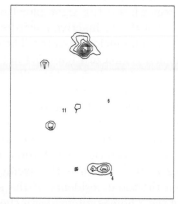

Figure 5.4 *Left: Locations of 552 leukemia cases in upstate New York, along with an approximate outline of the eight county study region. The rectangular region is* 1×1.2 *square units. Right: contour plot of the population density* $\lambda(t)$, *and the locations of the 11 hazardous waste sites.*

census blocks are available, but precise locations of the leukemia cases are not. Our methods require these precise locations, so we randomly dispersed the cases throughout their corresponding census blocks. An approximation to the census blocks was achieved using the Voronoi tesselation (Green and Sibson 1978) with the Euclidean distance metric: each point in the region was identified with the closest centroid. All the cases identified with a particular centroid were then randomly dispersed in the block identified with that centroid; if there was exactly one case in a block we placed it at the centroid. Five independent randomizations of the case locations (within census blocks) are used in our study; see the left panel of Figure 5.4 for one such example. This provides an ad-hoc sensitivity analysis for our approach. The approximate study region is also indicated in the left panel of Figure 5.4, but to avoid edge effects we consider the enlarged rectangular area (roughly 145×174 square kilometers) as the support for the landmarks. Our analysis is based on the Neyman–Scott model with the leukemia intensity rate specified by

$$\mu(t|\boldsymbol{x}) = \lambda(t)\left(\epsilon + \sum_{i=1}^{n(\boldsymbol{x})} h(t|x_i)\right),$$

where $\lambda(t)$ adjusts for population density, $h(t|x)$ is the Thomas intensity (5.5), and $\epsilon > 0$ is a background leukemia incidence rate. Our earlier treatment of the Neyman–Scott model extends without change to this form of the model because $\lambda(t)$ does not depend on x. We use a Strauss prior for the landmarks x. For $\lambda(t)$ we use a smoothed version of the population density based on the 1980 U.S. census (extrapolated to the large rectangular region); see the right panel of Figure 5.4. This plot also gives the locations of the 11 inactive hazardous waste sites suspected of causing elevated leukemia incidence rates. The original dimensions (kilometers) of the rectangular region in Figure 5.4 are divided by 145 for the purpose of this analysis, resulting in a rectangular region which is 1×1.2 square units. For the interaction radius in the Strauss prior we use $r = 0.1$ (approximately 14.5 kilometers), and the interaction parameter is taken as $\gamma_X = 0.1$. The choice of these parameter values is rather arbitrary, but recall from the toy example (Section 5.4.1) that varying these values seemed to have little effect on the posterior distribution. In the Neyman-Scott model we specify $\epsilon = 5.2 \times 10^{-4}$, which is the average leukemia incidence rate of 5.2 cases per ten thousand residents of the study region. For the standard deviation of the Thomas intensity we choose $\sigma = .01$ (approximately 1.45 km.). These hyperparameter values will remain fixed throughout our analysis of the New York leukemia data. In the context of the leukemia data and toxic waste sites the choice of the prior hyperparameters is of great importance. If we believe that the incidence locations are simply a function of population density, and are not influenced by any underlying sources (such as the waste sites), then the prior should place high probability on the empty configuration. This is certainly a reasonable (and perhaps conservative) assumption. Should the posterior indicate the existence of one or more landmarks (i.e. place low probability on the empty configuration), then one would have a strong argument in favor of clustering. It may also be reasonable to place high prior probability on the existence of one landmark, but to posit the existence of more than one landmark is probably stacking the deck a bit too much. It should be noted that the locations of the eleven toxic waste sites are not a part of our model. In our initial analysis we choose a prior which places high probability on the empty configuration: $\beta_X = 0.113$. Figure 5.5 is based on the data shown in the left panel of Figure 5.4, and shows the effect of varying the landmark weight κ^* on the number of posterior points produced. The number of landmarks given by the prior distribution is included for reference. For each figure we used a Silicon Graphics workstation to produce 1000 samples using the Kendall–Møller algorithm. (It took less than one minute to produce the samples from the prior, and just over 4.5 hours for $\kappa^* = 0.23\epsilon$.) It is clear from this figure that increasing the landmark weight has little effect on the number of posterior landmarks. In fact, the posterior probability of the empty configuration is approximately 80% for each value of κ^*, even though the prior probability of the empty configuration is about 50%.

Figure 5.6 is similar to Figure 5.5, except that we use $\beta_X = 0.5$. The

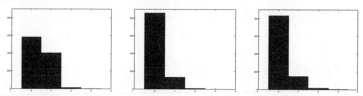

Figure 5.5 *Histograms of the number of points in the prior and two posterior distributions. Each plot is based on* 1000 *exact samples. The prior is Strauss, with* $r = 0.1$, $\beta_X = 0.113$ *and* $\gamma_X = 0.1$. *The likelihood is Neyman-Scott with a Thomas intensity,* $\epsilon = 5.2 \times 10^{-4}$, $\sigma = 0.01$. *Prior distribution; posterior with* $\kappa^* = 0.1\epsilon$; $\kappa^* = 0.23\epsilon$.

histogram for the prior shows that most of the prior weight is placed on configurations comprised of a single landmark. Nevertheless, the posterior distributions still strongly favor the empty configuration. As the landmark weight κ^* increases there is a slight movement towards configurations with at least one landmark. At large values of κ^*, limitations in computing power come into play. It took 146 hours to complete the simulation with $\kappa^* = 0.23\epsilon$. Initial runs with $\kappa^* = 0.25\epsilon$ indicate that the time required for a complete run is prohibitive, and the resulting histograms are virtually identical to that for $\kappa^* = 0.23\epsilon$. It is unfortunate that our exploration of this data is limited in this manner. Based on the results seen in the toy example of Section 5.4.1 and in the redwood seedling example of the next section, it seems likely that larger values of κ^* would result in a significant increase in the number of posterior landmarks. Figure 5.7 gives the

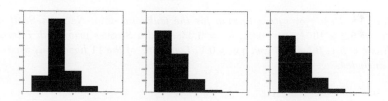

Figure 5.6 *Same as Figure 5.5 except that* $\beta_X = 0.5$.

posterior intensity contour plots when $\beta_X = 0.5$, for three values of κ^*. As κ^* increases, the posterior landmarks concentrate in two small areas (close to Syracuse and Binghamton) even though the *number* of posterior landmarks does not change significantly (see Figure 5.6). Figure 5.8 shows the relationship of the hazardous waste site locations to areas of high posterior landmark intensity. Note that several of the waste sites (1, 2, 3, and 4) are located close to areas of high intensity. To assess the significance of an

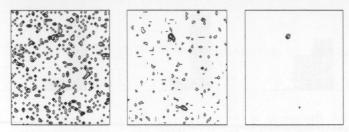

Figure 5.7 *Intensity contour plots for posterior distributions with* $\kappa^* = 0.1\epsilon$, $\kappa^* = 0.15\epsilon$, *and* $\kappa^* = 0.2\epsilon$. *Each plot is based on 1000 exact samples. The prior is Strauss, with interaction radius* $r = 0.1$, $\beta_X = 0.5$ *and* $\gamma_X = 0.1$. *The likelihood is Neyman-Scott with a Thomas intensity,* $\epsilon = 5.2 \times 10^{-4}$, $\sigma = 0.01$.

Figure 5.8 *Posterior intensity map for the leukemia data, Neyman–Scott model with* $\epsilon = 5.2 \times 10^{-4}$, $\sigma = 0.01$, $\kappa^* = 0.23\epsilon$, *and a Strauss prior with interaction radius* $r = 0.1$, $\beta_X = 0.5$ *and* $\gamma_X = 0.1$. *Locations of the 11 hazardous waste sites are included.*

elevated leukemia rate in the neighborhood of a given site, we compare the 'observed' with the 'expected' posterior landmark distribution. The relevant null hypothesis here is that the leukemia cases form an inhomogeneous Poisson process with intensity $\rho\lambda(t)$, where ρ is the average leukemia rate throughout the study region. To sample from the null distribution, we generate an artificial dataset using independent Poisson counts for each census tract, then analyze the artificial data in the same way as the original data. For the artificial data, histograms of the number of posterior landmarks (not shown) look almost identical to those in Figure 5.6, and intensity plots (also not shown) like those shown in Figure 5.7. In Figure 5.9 we compare the observed and expected posterior probabilities of at least one landmark within a given distance (0–7.25 kms) of site 1 (with the same hyperparameters as in Figure 5.6). The 20 dotted lines correspond to samples from the null distribution (20 different artificial datasets), and the 5 solid lines correspond to the data (five random dispersions of the leukemia cases throughout their corresponding census blocks). For this site, it is clear that increasing the landmark weight increases the disparity between the real and the simulated datasets. For $\kappa^* \geq 2.0$, these plots seem to provide some evidence of elevated leukemia rates in the neighborhood of site 1. (It should be noted that the vertical scales are different for each of the plots.) We re-

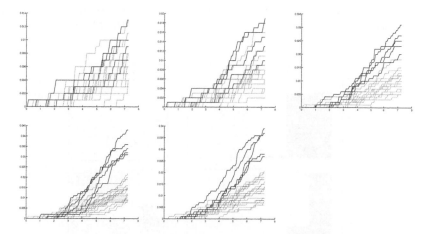

Figure 5.9 *Posterior observed (solid lines) and expected (dotted lines) probabilities of at least one landmark within a given distance (in kms.) of site 1. Neyman–Scott model with $\epsilon = 5.2 \times 10^{-4}$, $\sigma = 0.01$, and a Strauss prior with interaction radius $r = 0.1$, $\beta_X = 0.5$ and $\gamma_X = 0.1$. Top row: $\kappa^* = 0.1\epsilon$; $\kappa^* = 0.15\epsilon$; $\kappa^* = 0.2\epsilon$. Bottom row: $\kappa^* = 0.22\epsilon$; $\kappa^* = 0.23\epsilon$. Each plot is based on 1000 exact samples.*

peated the comparison shown in Figure 5.9 for each of the 11 hazardous waste sites, and found no evidence of elevated leukemia rates in the neighborhood of any site other than site 1. The complete set of comparisons can be seen in Loizeaux and McKeague (2001). Ghosh et al. (1999) found little evidence of elevated leukemia rates in the proximity of any of the waste sites. They applied a hierarchical Bayes generalized linear model to estimate the leukemia incidence rate locally, and included the hazardous waste site locations in their model (in particular, using the inverse distance to the nearest hazardous waste site as a covariate). Ahrens et al. (1999) found increased rates at three of the sites (1, 5 and 7). Denison and Holmes (2001) found significant evidence of increased leukemia rates in the neighborhoods of sites 1 and 2. Their Bayesian partition model assumes that the region can be split into local areas in which the leukemia counts come from the same distribution. The Voronoi tesselation (Green and Sibson 1978) is used to define these areas. Our results are mixed. On the one hand, the posterior distributions place approximately 50% of their mass on the empty configuration. But on the other hand, a comparison of the actual data with artificial datasets based only on the population density shows a marked difference at site 1.

5.6 Redwood Seedlings Data

These data have been analyzed by Strauss (1975), Ripley (1977), Diggle (1978) and van Lieshout (1995), among others and was also used in the previous chapter. Our results are shown in Figure 5.10. The prior hyperpa-

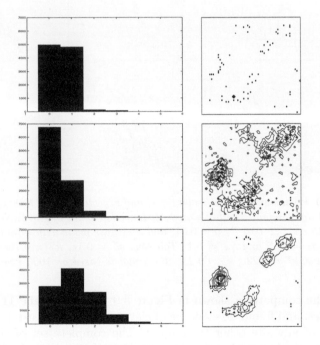

Figure 5.10 *Features of the prior and two posterior distributions for the redwood seedling data using the Neyman–Scott model, $\epsilon = 0.1$ and $\sigma = 0.07$. Top row: the Strauss prior, with interaction radius $r = 0.2$, $\beta_X = 0.124$, and $\gamma_X = 0.1$. Middle row: posterior with $\kappa^* = 0.5\epsilon$. Bottom row: posterior with $\kappa^* = \epsilon$. Left: histogram of the number of landmarks. Right: contour plot of the intensity. In each case 10,000 exact samples were drawn using the Kendall–Møller algorithm.*

rameters have been chosen to produce empty landmark configurations with probability 50%. For $\kappa^* = 0.5\epsilon$ the posterior shows a clear shift towards the empty configuration, but this tendency is reversed as κ^* is increased to ϵ. In addition, the intensity plots indicate that the posterior landmarks start to collect around the data as κ^* increases.

Acknowledgements

This research was partially supported by NSA Grant MDA904-99-1-0070 and NSF Grant 9971784. Equipment support was provided under ARO Grant DAAG55-98-1-0102 and NSF Grant 9871196.

CHAPTER 6

Bayesian Estimation and Segmentation of Spatial Point Processes Using Voronoi Tilings

S.D. Byers   *A.E. Raftery*

6.1 Introduction

In this chapter we consider Bayesian estimation and segmentation for non-homogeneous Poisson point processes in two dimensions. This work turns out to be easily generalizable to higher dimensions and to have similarities with work done in one dimension.

The principal motivation for the methods considered here can be found in Muise and Smith (1992), Dasgupta and Raftery (1998), Byers and Raftery (1998) and Allard and Fraley (1997). The simplest case is the segmentation of a point process of inhomogeneous rate, for instance the isolation of regions of high density in a point process in the plane, or in a higher dimensional space. Estimation of the rate is also an issue, one that might be intimately bound up with the segmentation as the segmentation methods might serve as a (semi-)parametric method of rate estimation.

In Dasgupta and Raftery (1998) the assumption is made that the regions of high density are formed by a mixture of Gaussian distributions on a background Poisson process and model-based clustering Banfield and Raftery (1993) is used to segment the data. This method is particularly suited to finding very linear features or those approximated by linear forms. In Byers and Raftery (1998) and Allard and Fraley (1997) the point process is assumed to be one of piecewise constant rate, with only two distinct rates being present. The related methods of kth nearest neighbors and Voronoi tile areas, respectively, are used to segment the data.

All of these methods provide some form of classification of the events in the process into high or low density regions. The aim of this paper is to explore methods of gaining more information, such as a region-based estimator with uncertainties and other complex posterior probabilities. The Voronoi tile area based method provides a region but this can be quite crude, and has no uncertainties. Fully Bayesian extensions to the Gaussian cluster approach with solutions via MCMC were considered in Bensmail et al. (1997). We intend to provide the analogous exploration for the approaches in Byers and Raftery (1998) and Allard and Fraley (1997).

Figure 6.1 shows some point process data from a minefield detection problem that might be segmented into two parts, high density and low density. The second panel shows the data that are estimated to be in the higher intensity region on the basis of their posterior probabilities. Section 6.5 discusses this example further and compares other methods of segmentation.

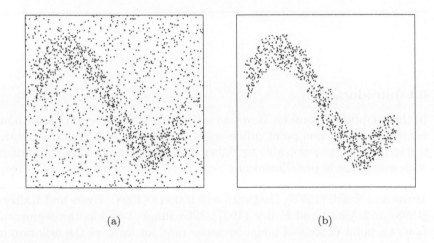

(a)   (b)

Figure 6.1 *Left panel shows a point process on a region, right panel shows an estimate of the points in the high rate region of the data. The detection rate is 97.0% while there are 4.2% false positives.*

The formulation used in the methods examined here is described in Section 6.2. Section 6.3 deals with the specifics of intensity estimation while Section 6.4 deals with aspects of segmentation. Some examples are shown in Section 6.5 while Section 6.6 contains some discussion of this and future work.

6.2 Proposed Solution Framework

6.2.1 Formulation

The proposed solution is to approximate a point process as a Poisson point process with piecewise constant rate. Thus the region under consideration, S, will be partitioned up into sets $B_k, k = 1, \cdots , K$ such that $\bigcup_{k=1}^{K} B_k = S$ and $B_i \cap B_j = \emptyset$, for all $i \neq j$. The sets will be generated and manipulated by constructing the Voronoi tiling of a set of synthetic generating points, in general unrelated to the data points themselves. Section 6.2.2

gives a brief account of Voronoi tiling. This framework will be treated in a Bayesian manner and solutions will be sought with Markov chain Monte Carlo methods.

In order to estimate the rate of a process as a function of space, each tile will have its own rate. The posterior distribution of numbers of tiles, tile positions and rates on each tile can then provide an estimate of the rate and the variability of the estimate as a function of space. A motivation for using this framework is that it has the potential to recover discontinuities in the rate of the process while approximating smooth gradients in intensity.

In order to segment a point process according to intensity the tiles will be grouped into two sets, each set having a different rate. More than two sets could be used in order to allow segmentation into more groups; indeed a variable number of sets could be used.

6.2.2 Voronoi Tilings

The Voronoi tiling of a set of points $\{c_k, k = 1, \cdots, K\}$ in the plane is a partition of the plane into regions $B_k, k = 1, \cdots, K$ such that for all $\mathbf{x} \in B_k$, $d(\mathbf{x}, c_k) < d(\mathbf{x}, c_l), l \neq k$. Voronoi tiles are always convex polygons. An overview of the history of Voronoi tilings with some applications and algorithms for computation can be found in Okabe et al. (2000). The partition generated is convenient in that it is easy to find which set, or tile, a point not in the generating set belongs to simply by determining to which generating point it is closest. This partition is also convenient in terms of its specification. Only the generating points need be specified to know the tiling, with the possible addition of a window defining the region under consideration. This framework is similar to that of the simple image segmentation example illustrating reversible jump Markov chain Monte Carlo in Green (1995). There a point process in one dimension has its rate estimated with possibilities for segmentation using a method that might be recast in a Voronoi tiling parameterization. Voronoi tilings extend to higher dimensions and algorithms and code exists to find them in general dimension.

6.2.3 Markov Chain Monte Carlo Using Dynamic Voronoi Tilings

In the model we use a piecewise constant rate point process which will be specified and controlled by the Voronoi tiling of a set of synthetic generating points. In the MCMC algorithms used to explore posterior distributions, simple changes in the tiling will enable movement around the state space. Local changes can be made to a tiling by moving, adding or deleting a generating point. Adding and deleting a point induces strictly local changes in terms of neighbor relations, but moving a generating point may theoretically induce larger scale changes, regardless of how small the movement is.

On the other hand small movements of generating points can induce much more subtle changes to the tiling than addition or deletion.

The task of computing Voronoi tilings we delegate to `deldir`, a package written by Rolf Turner at the University of New Brunswick. The driver routines in RATFOR were used both for our MCMC and via the Splus statistical package for display and interactive investigation purposes.

6.3 Intensity Estimation

6.3.1 Formulation

A general Bayesian formulation for the Voronoi tiling model of the intensity of a point process is:

$$
\begin{aligned}
\text{Likelihood} \quad &\propto \quad \prod_{k=1}^{K} \lambda_k^{n_k} \exp\{-\lambda_k a_k\} \\
\text{Priors} \quad & K \sim Poisson(\nu) \\
& \{\lambda_k; k = 1, \cdots, K\}|K \text{ iid } \sim Gamma(a, b) \\
& \{\, \mathbf{c}_k; k = 1, \cdots, K\}|K \text{ iid } \sim U\{S\}.
\end{aligned}
$$

Here the a_k are the areas of the Voronoi tiles, or the sets B_k, and the n_k are the numbers of points on each tile. Note that the number of tiles is not fixed in the formulation above, though it could be held at some fixed value. A hyperprior could be placed on the parameter ν, and ν could be updated in the MCMC schedule; this would allow greater flexibility in the ability to tile the region.

The posterior distribution induced by the above formulation is

$$
\pi(K, \{\lambda_k, \mathbf{c}_k; k = 1, \cdots, K\}|\mathbf{y}) \quad \propto \quad \frac{\nu^K}{K!} \prod_{k=1}^{K} \lambda_k^{n_k + a - 1} \exp\{-\lambda_k(a_k + b)\}.
$$

6.3.2 MCMC Implementation

In order to fit the models described here we use Markov chain Monte Carlo to simulate from the posterior distribution of the parameters given the data. The algorithms are of the Metropolis-Hastings type (Metropolis et al. 1953, Hastings 1970), which have as a special case the Gibbs sampler (Geman and Geman 1984). The samples obtained by these methods from the posterior distributions can then be used to form estimates of the intensity of the process or other quantities of interest.

Fixed Number of Tiles

First we consider the simple case where we use a fixed number of tiles in the partition. We first note the form of the full conditional distributions of the parameters:

$$\pi(\lambda_k|\ldots) \sim Gamma(n_k + a, a_k + b), \quad k = 1, \cdots, K$$

$$\pi(\mathbf{c}_k|\ldots) \propto \prod_{k=1}^{K} \lambda_k^{n_k+a-1} \exp\{-\sum_{k=1}^{K} \lambda_k(n_k + b)\}, \quad k = 1, \cdots, K.$$

The $|\cdots$ notation is used to denote conditioning on all other parameters and the data. In the full conditional for $\mathbf{c}_k$ the influence of the different values for the parameters is manifested implicitly in the values of a_k and n_k from the tiling induced by changes in $\mathbf{c}_k$.

The simplest sampler is one that employs a Gibbs step for each of the λ_k and then proposes moving one or more generating points in a Metropolis-Hastings step. A simple proposal for moving a generating point is uniform in a small square centered at the current position, rendering the move of the Metropolis type.

More general samplers might employ larger scale moves of generating points accompanied by the adjustment of the rate on the tile according to where the proposal generating point sits. It is possible to move more than one, or indeed all of the generating points. The following three schemes were implemented and used together;

- Propose movement of one generating point a small distance.

- Propose movement of a generating point to anywhere in the region. This is more like deleting a generating point and then adding a new one, see the next section. Also the intensity might be proposed to change.

- Propose to move all generating points by very small amounts. This will have the effect of facilitating capture of detail at discontinuities in the intensity.

Whichever of these proposals is tried, the basic elements of the move remain the same: propose movements of points, recompute the new tiling, find the acceptance probability for the new tiling versus the old tiling and update accordingly.

Figure 6.2 shows two possible methods of altering the tiling by movement or deletion of a generating point. In panel (a) the point to be moved is circled, its new position is shown with an empty circle. The changes to the tiling are shown with the dotted lines. In panel (b) a point to be deleted is circled.

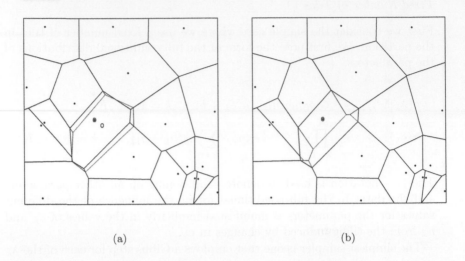

(a)   (b)

Figure 6.2 *Examples of two types of proposal for altering the Voronoi tiling in the MCMC algorithm. (a) Proposed movement of a generating point from the ringed point to the empty circle. (b) Deletion (or addition) of the ringed generating point. The dotted lines indicate the induced changes to the tiling.*

Variable Number of Tiles

It is desirable to consider a variable number of tiles in the tiling for reasons of flexibility and for mobility of the samplers. For instance, if too small a number of tiles is used, there might not be the ability to capture detail in the process. Regarding mobility, the addition or deletion of a tile can induce large plausible changes in the configuration. In order to sample from a target distribution with a variable number of variables, conventional MCMC must be augmented by reversible jump Markov chain Monte Carlo (Green 1995).

Similar moves will be made to those seen previously, but with the addition of a pair of moves that propose addition or deletion of a tile, with some suitable adjustment of the intensities on the tiles involved. An addition or a deletion has only local effects on the tiling but even small movements of tile generating points have the capability to change the tiling at long range. This has implications for computational efficiency, and the use of Voronoi tilings in Green (1995) exploits this with the use of the incremental tiling algorithm seen in Green and Sibson (1978).

We make the same type of addition and deletion steps as in the Voronoi example of Green (1995); the proposal distributions and acceptance probabilities are given there. Briefly, the position of a new generating point is sampled from the prior and the rate associated with the induced tile is proposed as $\lambda_{k+1}^* = \tilde{\lambda}_{k+1}v$ where $\tilde{\lambda}_{k+1}$ is the weighted geometric mean of the neighboring tiles and v is drawn from the distribution with density $f(v) = 5v^4/(1+v^5)^2$. The deletion move is the reverse of this construction.

6.4 Intensity Segmentation

The intention in this section is to segment point processes in the plane based on their intensities. Of particular interest is the case where a process is assumed to have two intensities. Thus the process has piecewise constant intensity, the region under consideration being partitioned into two disjoint sets, each having an intensity. The two sets need not form connected regions.

We consider the case of two regions, one of intensity λ_0, the other of intensity λ_1. The object is to partition the region of interest into two regions and make statements about the uncertainty with which this is done and the estimates of λ_0 and λ_1.

6.4.1 Formulation

Here S is the region under consideration, A_0 is the area of the sub-region assigned rate μ_0, in which N_0 data points fall, while A_1 is the area of the sub-region with N_1 data points which has rate $\mu_0 + \mu_1$. The rates are parameterized in this manner to give identifiability of the subregions; otherwise one might get swapping of the regions and the rates. In the minefield context these are a noise rate and a mine rate giving an intensity inside a minefield of $\mu_0 + \mu_1$ and outside of μ_0. The classification of the tiles will be coded in the variables d_k, indicating high or low rate. The positions of the tile generating points are $\{c_k : k = 1, \cdots, K\}$. The following formulation results from the above considerations:

$$
\begin{aligned}
\text{Likelihood} \quad &\propto \quad \mu_0^{N_0}(\mu_0 + \mu_1)^{N_1}\exp\{\mu_0 A_0 + (\mu_0 + \mu_1)A_1\} \\
\text{Priors} \quad &\quad \mu_0, \mu_1 \sim Gamma(a,b) \\
&\quad K \sim Poisson(\nu) \\
&\quad \{d_k; k = 1, \cdots, K\}|K \sim Bernoulli(p) \\
&\quad \{ c_k; k = 1, \cdots, K\}|K \sim U\{S\}.
\end{aligned}
$$

Fixed Number of Tiles

Here, as in the estimation context, it is possible to use a restricted set of models and fix the number of tiles used in tiling the region. In this case one sweep of the MCMC algorithm might look like the following:

- update the two rates of the process,

- change the classification of a tile from high to low, or vice versa,

- move a generating point, or even more than one.

The posterior distribution $\pi(\mu_0, \mu_1, \{d_k, \mathbf{c}_k; k = 1, \cdots, K\} | \mathbf{y})$ is proportional to

$$\mu_0^{N_0+a-1} \mu_1^{a-1} (\mu_0+\mu_1)^{N_1} \exp\{\mu_0(A_0+A_1+b)+\mu_1(A_1+b)\} \prod_{k=1}^{K} p^{d_k}(1-p)^{1-d_k}.$$

Thus, writing $\xi = \exp\{\mu_0(A_0 + A_1 + b) + \mu_1(A_1 + b)\}$ for notational convenience, the relevant full conditional distributions are given by

$$
\begin{aligned}
\pi(\mu_0, \mu_1 | \cdots) &\propto \mu_0^{N_0+a-1} \mu_1^{a-1} (\mu_0 + \mu_1)^{N_1} \times \xi \\
\pi(d_k | \cdots) &\propto p^{d_k}(1 - p)^{1-d_k} \mu_0^{N_0+a-1} \mu_1^{a-1} (\mu_0 + \mu_1)^{N_1} \times \xi \\
\pi(\mathbf{c}_k | \cdots) &\propto \mu_0^{N_0+a-1} \mu_1^{a-1} (\mu_0 + \mu_1)^{N_1} \times \xi.
\end{aligned}
$$

The MCMC steps for the segmentation are similar to those for the estimation algorithm, but additional care needs to be exercised in order to retain mobility. The rates of the process are updated together in a bivariate Hastings step as they may be negatively correlated. They are proposed by multiplying by $\exp(U)$, $U \sim \text{Uniform}(-d, d)$. This gives a simple proposal ratio of the ratio of the new to the old values, leading to an acceptance probability $\alpha = \min\{1, R\}$, where

$$
\begin{aligned}
R(\mu_0, \mu_1; \mu_0', \mu_1') &= \left(\frac{\mu_1' \mu_0'}{\mu_1 \mu_0}\right)^a \left(\frac{\mu_0'}{\mu_0}\right)^{N_0} \left(\frac{\mu_1' + \mu_0'}{\mu_1 + \mu_0}\right)^{N_1} \\
&\times \exp\{-A_0(\mu_0' - \mu_0) - (b + A_1)(\mu_1' + \mu_0' - \mu_1 - \mu_0)\}.
\end{aligned}
\tag{6.1}
$$

The d_k are proposed to their opposite value. This gives the acceptance ratio for $d_k = 0 \to d_k' = 1$ of

$$
R(d_k; d_k') = \left(\frac{\mu_0}{\mu_0 + \mu_1}\right)^{-n_k} \exp(-\mu_1 a_k),
\tag{6.2}
$$

where n_k is the number of points that fall on tile k and a_k is the area of tile k. The reverse move has acceptance ratio equal to the inverse of this.

A generating point's movement is proposed as in the estimation algorithm with the possible addition of a change in classification. The new tiling will define A_0' and A_1', while a new μ_0' and μ_1' may be proposed. Then the acceptance probability for the movement of a tile becomes $\alpha = \min\{1, R\}$, where

$$R(\mathbf{c}_k \quad ; \quad \mathbf{c}'_k) = (\mathrm{PR}) \left(\frac{\mu'_1 \mu'_0}{\mu_1 \mu_0} \right)^{a-1} \frac{\mu_0'^{N'_0}}{\mu_0^{N_0}} \frac{(\mu'_1 + \mu'_0)^{N'_1}}{(\mu_1 + \mu_0)^{N_1}} \tag{6.3}$$
$$\times \quad \exp\{ -A'_0 \mu'_0 + A_0 \mu_0 - (b + A'_1)(\mu'_1 + \mu'_0) + (b + A'_1)(\mu_1 + \mu_0) \}.$$

Here PR is the proposal ratio for any additional adjustments to the rates or the classifications. The proposal for the position (u'_k, v'_k) is symmetric so does not appear in the acceptance probability.

Variable Number of Tiles

As in the estimation formulation it is desirable to allow a variable number of tiles in constructing the partition of the region. The extension is similar to previously, it involves the addition of an addition and deletion pair of moves in the reversible jump MCMC framework.

The proposal for the position of a new generating point is uniform on the region S as before. Now the new induced tile must be assigned to either the high or low rate, the simplest choice being to generate this assignment randomly from the prior. The two rates do not have to be changed but some perturbation based on the change in the two areas or a random variable might be used. This leads to an acceptance ratio of the form,

$$R = \left(\frac{\mu'_0 \mu'_1}{\mu_0 \mu_1} \right)^{a-1} \frac{(\mu'_1 + \mu'_0)^{N'_1}}{(\mu_1 + \mu_0)^{N_1}} \frac{1}{p} \frac{\nu}{k+1} |J|$$
$$\times \exp\{ -b\mu'_0 A'_0 + (\mu'_0 + \mu'_1)A_1 - \mu_0 A_0 + (\mu_0 + \mu_1)A_1 \}.$$

Here J is the Jacobian for the transformation from μ_0, μ_1 to μ'_0, μ'_1, d_k. The factor of $1/p$ arises from the random assignment of d_k while the Poisson prior for K yields the next term. In this case it is hard to find a good proposal for addition of a point to the tiling.

6.5 Examples

6.5.1 Simulated Examples

A Sine Wave

The data shown in Figure 6.1(a) are from Byers and Raftery (1998). A region of higher density is bounded by two sine waves. We applied the segmentation method to these data to obtain the results in panel (b). The sinusoidal nature of the edges of the region is captured very well. The detection rate is 97.0%, with 4.1% false positives. This result was obtained with a fixed number of tiles, specifically 100, but we chose that to be more than we thought were needed. The add-delete steps had great difficulty being accepted. Greater detail in forming the proposals might improve the acceptance rate for jumps in the variable number of tiles model for these data, but it seems that most of the problem is inherent.

A Linear Feature

These data are from Dasgupta and Raftery (1998) and were used there to illustrate the mclust-em method. Figure 6.3(a) shows the data and Figure 6.3(b) shows a probability map for the minefield with the lightest color being zero and the darkest being one. This can be used to yield a segmentation of the data or information about the region in general.

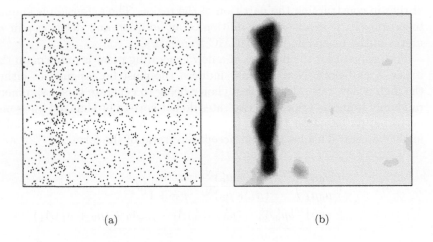

(a)   (b)

Figure 6.3 *(a) Simulated minefield with clutter. (b) Pointwise posterior probabilities of being inside the minefield.*

6.5.2 New Madrid Seismic Region

Data on earthquakes in the New Madrid seismic region are available from the Center for Earthquake Research and Information (CERI) web site at <http://samwise.ceri.memphis.edu>. We obtained these data and selected all earthquakes of magnitude greater than 2.5 from 1974 to 1992. To these data we applied the Bayesian intensity estimation and segmentation. The segmentation partitions the data into two regions, one near faults with many earthquakes and one away from major faults with lower, background intensity. The intensity estimation will provide detailed information regarding the intensity of earthquakes as a function of space over the region.

Figure 6.4(a) shows the earthquake locations. It is clear that the intensity varies in the region, in some places being very low and others quite high. Figure 6.4(b) shows the pointwise posterior probabilities being in a high-intensity area. In terms of detection of features the results are a linear section joined to a polygonal region with a separate area off on the right of the region.

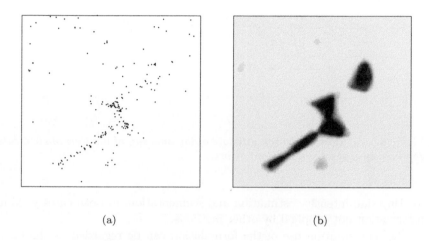

(a)   (b)

Figure 6.4 *(a) The New Madrid earthquake data, and (b) a Bayesian posterior probability segmentation.*

Figure 6.5(a) shows the posterior mean estimate of the intensity of occurrences in the region. The estimate shows inbuilt locally adaptive properties resulting from the flexible locally constant rate formulation. For example, the high intensity regions are captured with good detail and sharp edges. There is an apparent difference between this and the results of existing non-locally adaptive kernel smoothing methods. For example the above methods implemented in the Splus spatial statistics module (Diggle 1983) do not allow for a predominantly smooth estimate that also recovers the very localized regions of high intensity. They also do not provide variability estimates. Figure 6.5(b) shows the pointwise standard errors of the estimated intensity over the region.

6.6 Discussion

In this paper we have introduced methods for segmenting and estimating the rate of a spatial point process. They can represent local spatial homogeneity while allowing for discontinuities in the intensity. We use them

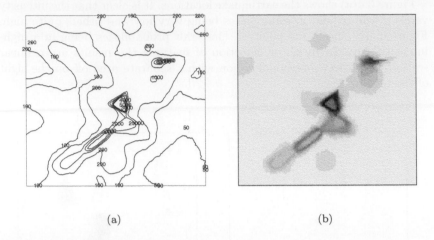

(a)　　　　　　　　　　　　　　　　　(b)

Figure 6.5 *(a) Posterior mean estimate of the intensity of the New Madrid data. (b) Pointwise posterior standard errors.*

for Bayesian intensity estimation and segmentation, in both cases yielding information not supplied by other methods.

The segmentation use of the formulation can be regarded as similar in nature to that of Allard and Fraley (1997), but providing a wealth of probabilistic information instead of only classifications. The method of Stanford and Raftery (1997), in its search for purely curvilinear features encounters problems for those in the form of an unstructured region, as evidenced in Figures 13 and 14 of their work. This is also the case in the S* parameterization of mclust-em used in Dasgupta and Raftery (1998). The method presented here is a density-based method, as opposed to density and geometry based, and so can capture (curvi)linear or bloblike features together with no additional specification.

Other methods have been proposed that address some aspects of the problems we are trying to solve here. The intensity of a spatial nonhomogeneous point process may be estimated by kernel or other nonparametric density estimation methods (Diggle 1985a, Scott 1992). Our proposal, which is also nonparametric, goes beyond this in that it provides formal probabilistic statements of uncertainty about the intensity, and provides an explicit model for segmentation and feature detection. Parametric models have also been used for intensity estimation in nonhomogeneous Poisson processes (Ogata and Katsura 1988); our approach here makes no parametric assumptions about the functional form of the intensity.

Often point process data come with some form of additional information in the form of marks. The formulation we are examining here is easily extendible to allow marks on the events in the process by use of a model for the marks. In the minefield example a multivariate mark or a more simple expression of confidence might be applicable. For example with each data point there might be a measurement $m_i \in (0,1)$. Real mines might have higher values while the noise points might have lower values. Marks in the minefield region could then be assumed distributed according to a mixture distribution $p f_1(w) + (1-p) f_2(w)$ while those outside would be just $f_2(w)$. This expresses the assumption that the data points in the minefield region are a mixture of noise and mines. Further work based on this formulation is under way.

PART II

Spatial process cluster modelling

CHAPTER 7

Partition Modelling

J.T.A.S. Ferreira   *D.G.T. Denison*   *C.C. Holmes*

7.1 Introduction

This chapter serves as an introduction to the use of partition models to estimate a spatial process $z(\boldsymbol{x})$ over some p-dimensional region of interest $\mathcal{X}$. Partition models can be useful modelling tools as, unlike standard spatial models (e.g. kriging) they allow the correlation structure between points to vary over the space of interest. Typically, the correlation between points is assumed to be a fixed function which is most likely to be parameterised by a few variables that can be estimated from the data (see, for example, Diggle et al. (1998)). Partition models avoid the need for pre-examination of the data to find a suitable correlation function to use. This removes the bias necessarily introduced by picking the correlation function and estimating its parameters using the same set of data.

Spatial clusters are, by their nature, regions which are not representative of the entire space of interest. Therefore it seems inappropriate to assume a stationary covariance structure over X. The partition model relaxes this assumption by breaking up the space into regions where the data are assumed to be generated independently from locally parameterised models. This can naturally place in a single region those points relating to an unusual cluster, and these points do not necessarily have to influence the response function in nearby locations. Further, by assuming independence between the regions the response function at the cluster centre tends not to be oversmoothed.

We now describe the partition model and its implementation in a Bayesian framework, via Markov chain Monte Carlo (MCMC) methods. We first give the model used for Gaussian response data but also discuss how the framework can be extended to count data (e.g. for disease mapping applications). Further, when analysing count data we show how, when covariate information is available, we can incorporate this into the analysis. This method is shown to be useful for both linear and nonlinear modelling of the covariate effects.

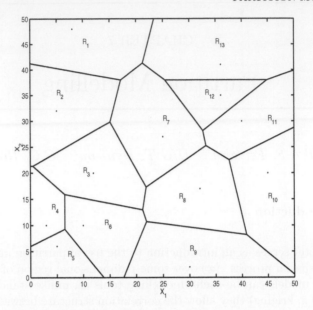

Figure 7.1 *An example of a Voronoi partition in two dimensions. The dots mark the centres of the regions, which are sometimes known as tiles.*

7.2 Partition Models

To be able to describe partition models for the special case of spatial data analysis we first need to define exactly what the term means. We use a definition similar to the original product partition models of Barry and Hartigan (1993) which is given in Chapter 7 of Denison et al. (2002b)

Definition *A partition model is made up of a number of disjoint regions $R_1, \ldots, R_k$ whose union is the domain of interest, so that $R_i \cap R_j = \emptyset$ for $i \neq j$ and $\bigcup_1^k R_i = \mathcal{X}$. The responses in each region, given parameters relating to each region $\boldsymbol{\phi} = (\phi_1, \ldots, \phi_k)$, are taken to be exchangeable and to come from the same class of distribution f.*

To illustrate the idea consider Fig. 7.1. Here we show a two-dimensional region split up by a partitioning method known as Voronoi tessellation which was introduced by Voronoi (1908) (for a modern view of Voronoi tessellations and their uses in spatial modelling see Okabe et al. (2000)). The dots mark the midpoint of each region (or tile) and are known as centres. Throughout this chapter we shall denote the centre of region i by $\boldsymbol{c}_i$ and write the collection of all centres as $\boldsymbol{c} = (\boldsymbol{c}_1, \ldots, \boldsymbol{c}_k)$.

Now, each region is defined as the portion of space closer to a particular centre than any other, with the lines giving the boundaries between regions. Hence, we can define the ith region by $R_i = \{\boldsymbol{x} \in \mathcal{X} : D(\boldsymbol{x}, \boldsymbol{c}_i) < D(\boldsymbol{x}, \boldsymbol{c}_j)$

for all $j \neq i\}$, where $D(\cdot, \cdot)$ is some user-specified distance metric. The Voronoi tessellation is completely defined by specifying a distance metric and the centres of the regions.

The choice of distance metric can be important, especially when there are many predictors, but as spatial data is usually low-dimensional (typically $p = 2$ or 3) the Euclidean distance, $D^2(\boldsymbol{x}_1, \boldsymbol{x}_2) = \sum_{i=1}^{p} (x_{1i} - x_{2i})^2$, is adequate (see Denison et al. (2002b) for more discussion). This is the metric we shall use throughout this chapter.

7.2.1 Partitioning for Spatial Data

A typical spatial dataset contains responses $y_1, \ldots, y_n$ at spatial locations $\boldsymbol{x}_1, \ldots, \boldsymbol{x}_n$. The responses are assumed to be observed only in the presence of some additive error component so that the data are generated by

$$y_i = \overline{y} + z(\boldsymbol{x}_i) + \epsilon_i,$$

for $i = 1, \ldots, n$, $\overline{y} = \frac{1}{n} \sum y_i$ and ϵ_i is a random draw from some zero-mean error distribution which we shall assume has variance σ^2 (the regression variance). Here $z(\boldsymbol{x})$ is some unknown spatial process which represents the deviations of the mean level of the process from the overall empirical mean, $\overline{y}$, for all $\boldsymbol{x} \in \mathcal{X}$. The aim of the analysis is to accurately approximate z given the observed dataset $\mathcal{D} = \{y_i, \boldsymbol{x}_i\}_1^n$. This generally involves some sort of smoothing of the observed data, which has a greater variance than the true process $z(\boldsymbol{x})$ if $\text{var}(\epsilon_i) > 0$.

Spatial smoothing, like smoothing in all regression contexts, involves a delicate trade-off between the bias and variance of the resulting estimate to the truth. We can see this by considering the mean-squared error between the estimator $\widetilde{z}$ and truth z, given by

$$
\begin{aligned}
MSE(\widetilde{z}, z) &= E\{(\widetilde{z} - z)^2\} \\
&= E\{(\widetilde{z} - E(\widetilde{z}))^2\} + \{E(\widetilde{z}) - z\}^2 \\
&= \text{var}(\widetilde{z}) + \text{bias}(\widetilde{z})^2.
\end{aligned}
$$

Here we see that to minimise MSE we must tradeoff fidelity in the mean-level (bias) with the variance of the estimator. Further, low variance estimators tend to have high bias and *vice versa*, for fixed MSE.

Now let us see how this relates to the partition model for spatial data, but first we shall discuss some of the common assumptions often made about spatial data. The first assumption nearly always made is that location matters. Thus, if a large response value is observed at $\boldsymbol{x}$, "nearby" points are more likely to also be large as their response values will be (positively) correlated with the value seen at $\boldsymbol{x}$. The strength of the correlation in the responses and the notion of "nearness" to $\boldsymbol{x}$ are both difficult to quantify. However, by parameterising the correlation structure and assuming it is independent of x (an isotropic correlation function) we can use the data to give us an idea of good values for the parameters in the correlation function.

Only considering stationary correlation functions, which do not vary with x, is restrictive. It does not allow for the possibility of discontinuities in $z(x)$, something that often makes sense in spatial data analysis, e.g. a break in the response surface might be due to a change from a plateau to a mountainous region. Hence, instead of trying to model the correlation structure directly, we choose to focus our attention on the response function, z.

Partition models give us a way of estimating z directly. Given a set of centres $c = (c_1, \ldots, c_k)$ we can define a tessellation of $\mathcal{X}$. Then, in each region R_i we can assume that the response function is constant with level μ_i. This gives rise to a particularly simple form for $\widetilde{z}$ with high bias but low variance, as the approximating function is so crude. However, we can decrease the bias in $\widetilde{z}$ by allowing the centres to be chosen by the data in an attempt to capture the underlying spatial variability in the truth, z. This is done with a view to placing more centres at locations with high variability and fewer where the function is relatively flat. Further reductions in bias can be achieved by allowing the estimate to be made up of a weighted sum of partition models, rather than just a single one, so that

$$\widetilde{z}(x) = \sum_{j=1}^{J} w^{(j)} \widetilde{z}^{(j)}(x | c^{(j)}), \tag{7.1}$$

where J is the number of models used to form the estimate, $w^{(j)}$ are the weights associated with each model such that $\sum_j w^{(j)} = 1$ and $\widetilde{z}^{(j)}$ is the estimate of the response function found with the jth set of centres $c^{(j)}$.

Various methods exist for producing suitable sets of estimating functions $\widetilde{z}^{(1)}, \ldots, \widetilde{z}^{(J)}$, most notably bootstrapping and Markov chain Monte Carlo (MCMC). Bootstrapping fits models on subsets of the data and then averages over those found, whereas MCMC relies on the setting up of a Markov chain designed so that its stationary distribution is the posterior distribution $p(c | \mathcal{D})$. In this paper we shall follow this second approach and adopt a Bayesian approach to making inference.

7.2.2 Bayesian Inference

Bayesian inference relies on Bayes Theorem

$$p(\boldsymbol{\theta} | \mathcal{D}) \propto p(\mathcal{D} | \boldsymbol{\theta}) p(\boldsymbol{\theta}),$$

to make inference about the distribution of some unknown vector of parameters, $\boldsymbol{\theta}$, in the light of observed data, $\mathcal{D}$. In the context of partition models the unknown parameters are the locations of the centres, $c = (c_1, \ldots, c_k)$, the levels in each region, $\boldsymbol{\mu} = (\mu_1, \ldots, \mu_k)$, as well as the regression variance σ^2. Note that the length of the vectors c and μ are of unknown, random length k.

To be able to determine the posterior distribution of the parameters we first need to assign prior distributions to all the unknowns. We do this hierarchically via the equality

$$p(\boldsymbol{\theta}) = p(k, c, \mu, \sigma^2) = p(\mu|c, \sigma^2)p(\sigma^2)p(c|k)p(k).$$

The prior we choose to place on the levels μ in each partition, given c and σ^2, is normal with mean zero and variance $\sigma^2 v$, for some prior constant v. The prior on the regression precision, σ^{-2}, is taken to be Gamma with parameters a and b. The prior over the positions of centres, assuming that knots can be placed anywhere in some space $\mathcal{T}$ (of the same dimension as $\mathcal{X}$), is taken to be uniform over $\mathcal{T}$. Finally we allow the number of centres to range from 1 to some prespecified maximum K. As we have little idea how many centres are required to adequately approximate the truth we take $k \sim U\{1, \ldots, K\}$, so the number of centres follows a discrete uniform distribution. Typically we take $K = n$. In summary

$$p(k) = \frac{1}{K}, \qquad k = 1, \ldots, K$$

$$p(c|k) = \frac{1}{\{\text{Area}(\mathcal{T})\}^k}, \qquad c \subset \mathcal{T}^k$$

$$p(\sigma^{-2}) = Gamma(\sigma^{-2}|a, b), \qquad \sigma^2 > 0$$

$$p(\mu|c, \sigma^2) = \prod_1^k N(\mu_i|0, \sigma^2 v), \qquad \mu \in \mathbb{R}^k.$$

This completes the prior specification on the unknowns in the partition model. In reality, as k is a deterministic function of c (i.e. just the number of elements in c) it is equivalent to remove the prior on k and just think of a prior on the centres such that

$$p(c) = \frac{1}{K}\frac{1}{\{\text{Area}(\mathcal{T})\}^k}, \qquad c \in \bigcup_{k=1}^K \mathcal{T}^k.$$

This simplifies the notation and is the prior we shall use for the rest of the paper.

Now, given the model parameters $\boldsymbol{\theta} = (c, \mu, \sigma^2)$, we can write down the likelihood of the data, $p(\mathcal{D}|\boldsymbol{\theta})$, after assuming a specific form for the error distribution. For the time being we shall assume that each $\epsilon_i \sim N(0, \sigma^2)$ and that they are all mutually independent. Hence, we find that

$$p(\mathcal{D}|\boldsymbol{\theta}) = (2\pi\sigma^2)^{-n/2}\exp\left\{-\frac{1}{2\sigma^2}\sum_1^n(y_i - \bar{y} - \mu_{r(i)})^2\right\},$$

for $k = 1, \ldots, K$ and zero otherwise. Here $r(i)$ is the index of the region that the ith response is located in.

The priors we chose to take over the levels, $\boldsymbol{\mu}$, and the regression variance, σ^2, aid our model in an important way. They are known as conjugate priors as they allow us to analytically determine the marginal likelihood

$$
\begin{aligned}
p(\mathcal{D}|\boldsymbol{c}) &= \int_0^\infty \int_{\mathbb{R}^k} p(\mathcal{D}|\boldsymbol{\mu}, \sigma^2, \boldsymbol{c}) p(\boldsymbol{\mu}|\sigma^2, \boldsymbol{c}) p(\sigma^2) d\boldsymbol{\mu} d\sigma^2 \\
&= \frac{(2b)^a \Gamma(a^\star)}{\pi^{n/2} \Gamma(a)} \frac{(2b^\star)^{-a^\star}}{\prod_1^k (n_i + 1)^{1/2}},
\end{aligned}
\tag{7.2}
$$

for $k = 1, \ldots, K$, and zero otherwise. Also, $a^\star = a + n/2$ with

$$
b^\star = b + \frac{1}{2} \sum_{i=1}^k \left\{ \sum_{j:r(j)=i} (y_j - \overline{y})^2 - \frac{v n_i^2}{1 + v n_i} \overline{y}_i^2 \right\},
$$

and n_i is the number of points in region i and $\overline{y}_i$ is the mean of those points in region i, i.e. $n_i = \sum_{j:r(j)=i} 1$ and $\overline{y}_i = n_i^{-1} \sum_{j:r(j)=i} (y_i - \overline{y})$.

The importance of using conjugate priors is that it allows us to reduce the dimension of the posterior from which we wish to sample from as now we only need to determine

$$
p(\boldsymbol{c}|\mathcal{D}) \propto p(\mathcal{D}|\boldsymbol{c}) p(\boldsymbol{c}).
\tag{7.3}
$$

Thus, the unknowns remaining are only the number of centres and their locations. The dimension of this posterior is under half the dimension of all the unknowns, $\boldsymbol{\theta} = (\boldsymbol{c}, \boldsymbol{\mu}, \sigma^2)$.

Note that, conditional on $\boldsymbol{\theta}$, the partition model for Gaussian data can be written as a linear model

$$
\boldsymbol{Y} = \boldsymbol{B}\boldsymbol{\mu} + \boldsymbol{\epsilon},
$$

where $\boldsymbol{Y} = (y_1 - \overline{y}, \ldots, y_n - \overline{y})'$, $\boldsymbol{\mu}$ and $\boldsymbol{\epsilon}$ are column vectors of the levels and errors, respectively. Lastly, $\boldsymbol{B}$ is known as the basis function matrix of dimension $n \times k$ where each row contains a one in position $r(i)$ and zeros elsewhere, i.e. $B_{ij} = 1$ if $j = r(i)$ and zero otherwise. Note that the form of this matrix is a deterministic function of the locations of the centres. In this setup we know, from standard Bayesian linear model results (e.g. Bernardo and Smith (1994), Denison et al. (2002b)), that the fitted values of all the datapoints given $\boldsymbol{c}$ and k are $\widetilde{\boldsymbol{Y}} = \{\widetilde{z}(\boldsymbol{x}_1|\boldsymbol{c}), \ldots, \widetilde{z}(\boldsymbol{x}_n|\boldsymbol{c})\}'$, where given by

$$
\widetilde{\boldsymbol{Y}} = E(Y|\mathcal{D}, \boldsymbol{c}) = \boldsymbol{B}(\boldsymbol{B}'\boldsymbol{B} + \boldsymbol{V}^{-1})^{-1}\boldsymbol{B}'\boldsymbol{Y},
\tag{7.4}
$$

where $\boldsymbol{V} = v\boldsymbol{I}_k$, where $\boldsymbol{I}_k$ is the k-dimensional identity matrix. When writing $\boldsymbol{S} = \boldsymbol{B}(\boldsymbol{B}'\boldsymbol{B} + \boldsymbol{V}^{-1})^{-1}\boldsymbol{B}'$, so that $\widetilde{\boldsymbol{Y}} = \boldsymbol{S}\boldsymbol{Y}$, we refer to $\boldsymbol{S}$ as the smoothing matrix. The ith row of $\boldsymbol{S}$ gives the weight that each observed response has on the ith fitted value $\widetilde{z}(\boldsymbol{x}_i|\boldsymbol{c})$. Thus, row i gives the equivalent kernel (Hastie and Tibshirani 1990) at the i observed location, $\boldsymbol{x}_i$. This allows us to determine the datapoints that have a significant effect on the estimated response at $\boldsymbol{x}_i$. We shall give examples of equivalent kernels later on in Section 7.2.5 when we discuss the kernels induced by the prior we have chosen for the partition model.

7.2.3 Predictive Inference

Given all the unknown parameters, $\boldsymbol{\theta}$, from normal theory we know that the posterior predictive distribution at a general point $\boldsymbol{x} \in \mathcal{X}$ is given by

$$p(y|\boldsymbol{x}, \boldsymbol{\theta}, \mathcal{D}) = N(y|\mu_{\boldsymbol{x}}, \sigma^2), \tag{7.5}$$

where $\mu_{\boldsymbol{x}}$ is the level of the region that $\boldsymbol{x}$ lies in. Further, as we used conjugate priors for $\boldsymbol{\mu}$ and σ^2 we can integrate them with respect to their posterior distributions (e.g. see Denison et al. (2002b)) from (7.5) and we find that

$$p(y|\boldsymbol{x}, \boldsymbol{c}, \mathcal{D}) = Student\left\{\frac{n_{\boldsymbol{x}}\overline{y}_{\boldsymbol{x}}}{n_{\boldsymbol{x}} + v^{-1}}, b^{\star}\left(1 + \frac{1}{n_{\boldsymbol{x}} + v^{-1}}\right), a^{\star}\right\}, \tag{7.6}$$

where $n_{\boldsymbol{x}}$ and $\overline{y}_{\boldsymbol{x}}$ are defined similarly to $\mu_{\boldsymbol{x}}$.

However, to obtain the posterior predictive distribution of the response at location $\boldsymbol{x}$ we must marginalise further over the posterior distribution of the centres and the number of them. That is, we need to determine

$$p(y|\boldsymbol{x}, \mathcal{D}) = \int p(y|\boldsymbol{x}, \boldsymbol{c}, \mathcal{D})p(\boldsymbol{c}|\mathcal{D})d\boldsymbol{c}. \tag{7.7}$$

Generally $p(\boldsymbol{c}|\mathcal{D})$ is not available in a computationally tractable form so we need to approximate it. As we have mentioned before, Markov chain Monte Carlo methods can be used to simulate from this distribution and in the next section we will go into more details about this. In summary, an approximate sample of models (or vectors of centre locations) $\boldsymbol{c}^{(1)}, \ldots, \boldsymbol{c}^{(J)}$ is generated from $p(\boldsymbol{c}|\mathcal{D})$ and the true posterior is approximated by a discrete uniform distribution on this sample of J models. Each of these models gives rise to a separate model for the responses at $\boldsymbol{x}$ as $z^{(j)}(\boldsymbol{x})$ is given by

$$\widetilde{z}^{(j)}(\boldsymbol{x}) = E(y|\boldsymbol{c}^{(j)}, \mathcal{D}) = \frac{n_{\boldsymbol{x}}\overline{y}_{\boldsymbol{x}}}{n_{\boldsymbol{x}} + v^{-1}}$$

Hence, the approximation to $p(y|\boldsymbol{x}, \mathcal{D})$ is a mixture of Student distributions and the overall estimate to the true spatial process, $z(\boldsymbol{x})$, is

$$\widetilde{z}(\boldsymbol{x}) = \frac{1}{J}\sum_{j=1}^{J}\widetilde{z}^{(j)}(\boldsymbol{x}).$$

7.2.4 Markov Chain Monte Carlo Simulation

We now describe how to generate samples of models $c^{(1)}, \ldots, c^{(J)}$ from the posterior of interest for partition models. We use a Markov chain Monte Carlo sampler (e.g. Gilks et al. (1996), Robert and Casella (1999)) which constructs a Markov chain whose stationary distribution is $p(c|\mathcal{D})$, the target posterior. This is achieved using an algorithm which makes transitions from the current set of model parameters according to a known set of rules. The most important of these are that the chain is irreducible and aperiodic. In addition, when sampling from a parameter vector with random dimension, as we are here, we also need the chain to be reversible (see Green (1995)).

We now give the basic algorithm for sampling from the partition model. Note that we shall refer to the comments included in the algorithm later on in Section 7.4.2.

Reversible jump sampler

Set $t = 0$ and initial values of parameters to $c^{(0)}$;

REPEAT

/*Draw auxiliary variables here*/

Set u_1 to a draw from a $U(0, 1)$ distribution;

IF $u_1 \leq b_k$

$c' = $ **Birth-proposal**$(c^{(t)})$;

ELSE IF $b_k < u_1 \leq b_k + d_k$

$c' = $ **Death-proposal**$(c^{(t)})$;

ELSE

$c' = $ **Move-proposal**$(c^{(t)})$;

ENDIF;

Set u_2 to a draw from a $U(0, 1)$ distribution;

IF $u_2 < \min\{1, BF(c', c^{(t)})\}$ /*ACCEPT NEW MODEL?*/

$c^{(t+1)} = c'$;

ELSE

$c^{(t+1)} = c^{(t)}$;

ENDIF;

/*Sample other parameters now*/

$t = t + 1$;

Take every mth value of $c^{(t)}$ after initial burn-in to be in sample;

END REPEAT;

In the above algorithm b_k and d_k are the probabilities of attempting a *BIRTH* and *DEATH* step when there are currently k centres. The rest of the probability $(1 - d_k - b_k)$ is reserved for the *MOVE* step. We choose to take $b_k = d_k = m_k = \frac{1}{3}$ for all k.

The *BIRTH* step proposes to increase the number of centres by one while the *DEATH* step does the reverse. The *MOVE* step proposes a change in location of one of the centres currently in the model. These proposed models are generated by their conditional priors as described in Denison et al. (2002b) so that the acceptance probability is the minimum of one and the Bayes factor in favour of the proposed model over the current one, i.e.

$$\min\left\{1, BF(\boldsymbol{c}', \boldsymbol{c}^{(t)})\right\} = \min\left\{1, \frac{p(\mathcal{D}|\boldsymbol{c}')}{p(\mathcal{D}|\boldsymbol{c}^{(t)})}\right\},$$

(which can be calculated using (7.2)). This particularly simple form of the acceptance probability results from the fact that we have chosen to control the possible values of k ($1 \leq k \leq K$) through the definition of the likelihood (7.2). This is in contrast to the usual approach that alters b_k, d_k and m_k for values of k equal to 1 and K.

In summary, after the initial setting of the model parameters the algorithm proceeds by making random choices between proposing to add, delete or remove a centre from the current tessellation. If the marginal likelihood of the proposed model, $\boldsymbol{c}'$, is greater than that for the current model, $\boldsymbol{c}^{(t)}$, then the proposed model is accepted. If this is not the case then the proposed model is accepted with the Bayes factor. After some number of burn-in iterations every mth model is taken to be in the generated sample. In the terminology of optimisation the MCMC algorithm performs a stochastic hill-climb, but because of the special way we make transitions we can use its output to approximate integrals, rather than just to identify local modes.

7.2.5 Partition Model Prior

Although we have described how to use MCMC techniques to draw samples from the model posterior, $p(\boldsymbol{c}|\mathcal{D})$, it is also of interest to see the effect that the prior $p(\boldsymbol{c})$ has on the inference. To do this we generate samples from $p(\boldsymbol{c})$, using the MCMC algorithm, which allows us to approximate the prior predictive distribution

$$p(y|\boldsymbol{x}) = \int p(y|\boldsymbol{x}, \boldsymbol{c})p(\boldsymbol{c})d\boldsymbol{c},$$

for all values $\boldsymbol{x} \in \mathcal{X}$. In addition, as we can think of the partition model as a Bayesian linear model given $\boldsymbol{c}$, we can look at the equivalent kernels (Hastie and Tibshirani 1990) at different locations $\boldsymbol{x} \in \mathcal{X}$. These allow us to see the degree of smoothing that the model induces at different points in the region of interest. For a location x and generated sample $c^{(1)}, \ldots, c^{(J)}$, the equivalent kernel is given by

$$K(x, x') = \frac{1}{J} \sum_{j=1}^{J} \frac{\chi(r_{x'}^{(j)} = r_x^{(j)})}{\text{Area}^{(j)}(r_x^{(j)})}, \qquad x' \in \mathcal{X},$$

where $r_x^{(j)}$ is the index of the region that x belongs to with configuration $c^{(j)}$, similarly for $r_{x'}^{(j)}$, and $\text{Area}^{(j)}(i)$ is the area of the ith region in the jth configuration. Also, $\chi(\cdot)$ is the indicator function that takes the value one when the statement is true and zero otherwise.

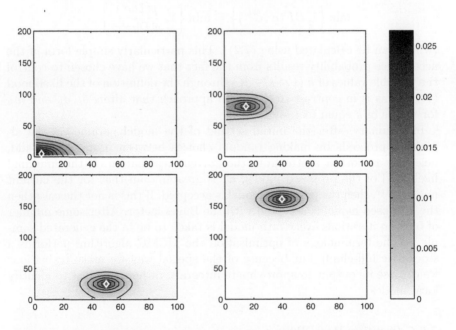

Figure 7.2 *The equivalent kernels produced by the partition model prior at (5,5), (15,80), (50,25) and (40,160). The points at which the kernels have been determined are shown by the diamonds.*

To allow us to easily display the equivalent kernels at various values of x we can discretise the region of interest into a regular grid of points and approximate the areas required by the number of points that lie in them. In this way we produced Fig. 7.2 using a 40×40 grid and taking $\mathcal{X} = [0, 100] \times [0, 200]$. Note how the prior induces spherical kernels (except for the kernel for the boundary point) that are very similar to independent normal kernels. Normal kernels have been used extensively before in spatial data analysis so the partition model prior appears acceptable. Note that the priors are not elliptical because of the different scales on each axis.

7.3 Piazza Road Dataset

To illustrate the performance of the method on data where the response can be assumed Gaussian we use the Piazza Road dataset studied in Higdon et al. (1999). The dataset is taken from an environmental monitoring study into the spatial distribution of dioxin concentrations on the Piazza Road, a part of an EPA Superfund site in Missouri. The dataset was introduced by Higdon et al. (1999) as an example of one in which it is expected that the spatial dependence between points varies with location. The response of the data was taken to be the natural logarithm of the dioxin concentration at the 600 spatial locations given.

We ran the MCMC simulation algorithm for 100,000 burn-in iterations and used 1 million further iterations to obtain the approximate sample of models, choosing every 100th to be in the generated sample. We used vague priors for the other parameters that we need to set, taking $a = b = 0.01$ for the Gamma prior on the regression variance and $v = 10$ for the prior variance of the levels. This value of v was chosen as the range of the data was around 6 and we wished that all the possible posterior values of the levels, μ_i, were contained within one standard deviation of the mean of their prior distribution.

In the top-left plot of Fig. 7.3 we display the estimated log concentration found using the partition model over a 40×40 grid. We see how the model allows for sharp differences between the mean level at points close together in space (the top-left portion of the plot) and also smooths well in regions with little spatial variability (the bottom half of the plots).

The other three plots give the equivalent kernels for three of the points from Fig. 7.2. We notice now that none of these three posterior kernels approximates a radial relationship between distance and correlation. For example, at (15,80) the points just below it are hardly used to estimate the mean value at (15,80). The three kernels are of very different shape, one horizontal, one vertical and one covers a large area. This highlights how the partition model can adapt to the underlying response surface to allow different degrees of smoothing at different locations, as well as the direction in which the most similar points lie.

7.4 Spatial Count Data

Although the observations in many spatial applications can be thought of as being generated from some true process together with Gaussian additive errors, this is not always the case. One important such application is disease mapping (e.g. Clayton and Kaldor (1987), Elliott et al. (1999), Lawson (2001)) where the observations are counts of disease. As counts are, by their very nature, integer-valued it is inappropriate to assume Gaussian errors. Even though by taking the square-root of Poisson observations we

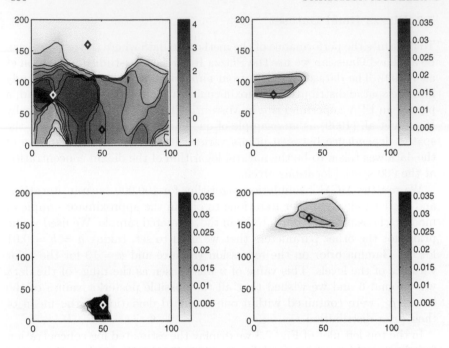

Figure 7.3 *The top-left figure is the surface fitted by the partition model to the logarithm of the dioxin concentrations in the Piazza Road dataset. The other three figures are the equivalent kernels produced by the partition model at the points (15,80) (top-right), (50,25) (bottom-left) and (40,160) (bottom-right). The diamonds give the positions of these points and are also displayed in the top-left figure (the colour of the diamonds is unimportant and was chosen to ensure that they were visible over the background surface).*

obtain more normal-looking response data (see, for example, Clyde (1999)) we would prefer to model the counts directly but still use the partition model framework. One way to do this is to follow the ideas in Denison and Holmes (2001). Here a Bayesian partition model was used to obtain a map of the disease risk for a leukaemia dataset. The difference with the earlier work we have seen is that the observations were modelled with a Poisson, rather than a Gaussian, distribution.

One limitation of the work of Denison and Holmes (2001) is that it does not allow for the incorporation of covariates into the model. However, in disease mapping applications, the main interest of the study is often the relationship between the covariates and the relative risk. So, in this chapter, we extend the Bayesian partition model for disease mapping to allow for covariates with the spatial process being used to "soaking up" the variation due to unobserved covariates. Before we get onto how to incorporate covariates we first review the Poisson-Gamma model for disease mapping as outlined in Denison and Holmes (2001).

7.4.1 The Poisson-Gamma Model for Disease Mapping

In a disease mapping context the spatial process under consideration is the disease risk at every location $x \in \mathcal{X}$. Alternatively, we can model the relative risk which is more conveniently calibrated so that values greater than one relate to more than average risk and *vice versa* for values less than one. The data is usually available in the form $\mathcal{D} = (y_i, N_i, x_i)_{i=1}^{n}$, where y_i is the observed number of disease cases in the ith enumeration district (ED) (e.g. county, electoral ward), N_i is the number of susceptibles in the same district and x_i is the centroid of that ED. Then, if we denote by ξ_0 the observed mean risk, so that $\xi_0 = \sum_1^n y_i / (\sum_1^n N_i)$, the expected risk at any point in the ith ED is $E_i = \xi_0 N_i$.

Again we suggest using the partition model to produce estimates to the relative risk, $\xi(x)$, at all values $x \in \mathcal{X}$. For rare diseases the Poisson model for incidence can be adopted for which we assume that each observation, y_i, is generated according to

$$p(y_i | x_i, N_i, c, \mu) = Poisson\{E_i \xi(x_i)\},$$

where $\xi(x_i) = \mu_{r(i)}$, the relative disease level associated with the region of the tessellation that x_i resides in.

As before, if we use conjugate priors for the levels we find that inference is considerably simplified and reduces to just determining $p(c|\mathcal{D})$. This can be done by taking the prior for the levels to be independent Gamma(γ_1, γ_2) distributions, i.e.

$$p(\mu_i) = \frac{(\gamma_2)^{\gamma_1}}{\Gamma(\gamma_1)} \mu_i^{\gamma_1 - 1} \exp(-\gamma_2 \mu_i), \quad \mu_i > 0,$$

for $i = 1, \ldots, k$. Using results relating to the conjugate Poisson-Gamma model we can again determine the marginal likelihood of the data given just the locations of the centres, c. We find that the logarithm of the marginal likelihood is given by

$$\log p(\mathcal{D}|c) = \text{constant} + \sum_{i=1}^{k} \left\{ \log \Gamma(\gamma_1 + n_i \bar{y}_i) - (\gamma_1 + n_i \bar{y}_i) \log(\gamma_2 + n_i \bar{N}_i) \right\},$$

where again n_i is the number of points in the ith region and $\bar{y}_i$ is the mean of the observations in that tile. Similarly, $\bar{N}_i$ is the mean number of susceptibles in the ith tile.

The posterior predictive density, $p(y|x, c, \mathcal{D})$ is also available in closed form and follows a Poisson-Gamma distribution (see Chapter 7, Denison et al. (2002b)). Also note that similar analytic tractability is also found for diseases that are not rare when we use the Beta-Binomial model for the disease risk.

The prior on c can be chosen in an identical way to the Gaussian case, however, we also need to set the γ_1 and γ_2 parameters. We have a lot of information about the range of values we expect for the relative risk and we need to incorporate this in our choice for these parameters. For instance, the mean relative risk should be one (with no data we would expect the relative risk in each area to be around one) which leads us to setting $\gamma_1 = \gamma_2$. This leaves us only one degree of freedom in our prior choice.

A sensible assumption is that the relative risk in any region is more likely to be within 5 times the mean relative risk (one), than outside this range. Thus, any value of γ_1 for which

$$\int_{0.2}^{5} \frac{(\gamma_1)^{\gamma_1}}{\Gamma(\gamma_1)} \mu^{\gamma_1 - 1} \exp(-\gamma_1 \mu) d\mu \; > 0.5,$$

appears reasonable. This leads us to choosing values of $c > 0.317$. Denison and Holmes (2001) assign a uniform prior to the logarithm of c and sample it from the data and to minimise the number of user-set parameters.

This model is well-suited to estimating the disease risk over an area of interest when there are no covariates, but problems arise when they are present in the data. In this case the usual way to include covariates is through an exponential transform so that the relative risk surface is given by

$$\xi(\boldsymbol{x}, \boldsymbol{x}^c) = \mu_{\boldsymbol{x}} \exp(\beta_1 x_1^c + \ldots + \beta_q x_q^c),$$

with $\boldsymbol{x}$ a general location, as before, and $\boldsymbol{x}^c = (x_1^c, \ldots, x_q^c)$ is a general vector of the q covariate values, measured at $\boldsymbol{x}$. Technical difficulties arise with this setup as there is now no conjugate prior for the coefficients $\boldsymbol{\beta} = (\beta_1, \ldots, \beta_q)$, so we cannot integrate them out. Instead we have to sample them to make posterior inference and in the next section we detail how we tackle this problem.

7.4.2 Disease Mapping with Covariates

Efficient sampling methods for Bayesian generalized linear models are well-known (see Dey et al. (2000) for an overview) so the first step to incorporating covariates is to rewrite the model in this form. First of all, we define some notation. Let $\boldsymbol{d_x}$ be a k-dimensional column vector with $\boldsymbol{d}$ having zeros everywhere except the component with the same index as the region that $\boldsymbol{x}$ is located in. Also take $\boldsymbol{\delta} = (\log \mu_1, \ldots, \log \mu_k)'$ so that $\boldsymbol{d_x'} \boldsymbol{\delta} = \log \mu_{\boldsymbol{x}}$, the logarithm of the level associated with location $\boldsymbol{x}$. We can interpret $\boldsymbol{d}$ as a vector gives us the index of the region that $\boldsymbol{x}$ lies in and $\boldsymbol{\delta}$ giving us the logarithm of the level in each of those regions. Now take $\boldsymbol{\beta} = (\beta_1, \ldots, \beta_p)'$ and let $\boldsymbol{b} = (x_1^c, \ldots, x_q^c)'$. Here, $\boldsymbol{\beta}$ is the vector of covariate coefficients and $\boldsymbol{b}$ gives the covariate values associated with the general datapoint at x. Hence, we can write

$$\log \xi(x, z) = d'_x \delta + b'_x \beta, \qquad (7.8)$$

and we assume that $y \sim Poisson(E_i \xi(x, z))$. Using the terminology of generalized linear models, we call the right hand side of (7.8) the linear predictor.

We wish to include a residual effect in (7.8) which is used to capture extra-Poisson variation often present in disease mapping datasets with low number of counts. The residual component can be thought of as representing random effects that are not captured by the covariates or the spatial model. The residual effects form auxiliary variables which augment the space over which we have to perform the sampling but greatly ease the computational convenience of the algorithm. However, to take advantage of this method we choose to place a prior over the logarithm of the levels in each region, δ, rather than levels themselves, μ, as before.

The priors that we adopt are

$$p(\delta, \beta | \sigma^{-2}) p(\sigma^{-2}) = N(0, \sigma^{-2} V) Gamma(\tau_1, \tau_2) \qquad (7.9)$$
$$p(\eta | \delta, \beta, \sigma^{-2}) = N(B[\delta \ \beta]', \sigma^2 I_n) \qquad (7.10)$$

where V is a known constant matrix, B is the design matrix whose ith row is $(d_{x_i}, b_{x_i})'$, and $\eta = (\eta_1, \ldots, \eta_n)$ and σ^2 are the auxiliary variables we have chosen to introduce. With these extra variables we can write

$$\eta = B \begin{pmatrix} \delta \\ \beta \end{pmatrix} + \epsilon,$$

where $\epsilon \sim N(0, \sigma^2 I)$. So, by comparison with a Bayesian linear model (Lindley and Smith 1972), we see that we can think of the η as a vector of responses observed with additive independent Gaussian errors. Hence, the form of the posterior distribution of δ and β given η is known, and we can even find the marginal likelihood of the data given η and the centres, c. We find that

$$p(\mathcal{D}|c, k, \eta) \propto \frac{|V^\star|^{1/2}}{|V|^{1/2}} (\tau_2^\star)^{-\tau_1^\star}, \qquad (7.11)$$

where $\tau_1^\star = \tau_1 + n/2$ and

$$\tau_2^\star = \tau_1 + \frac{1}{2} \{\eta' \eta - (m^\star)'(V^\star)^{-1} m^\star\}$$
$$V^\star = (B'B + V^{-1})^{-1}$$
$$m^\star = V^\star B' \eta.$$

Given the auxiliary variables η, everything else is simple to update as we can use the standard algorithm given before for the Gaussian case. We can propose to add, delete and move a centre and accept the proposed model with the ratio of marginal likelihoods, which can be found using

(7.11). However, one difference is that we must store current values of the coefficients and regression precision to allow us to update $\boldsymbol{\eta}$. Hence, at the second comment in the algorithm, we draw them from their conditional posterior using

$$
\begin{aligned}
p(\sigma^2|c, \boldsymbol{\eta}, \mathcal{D}) &= Gamma(\tau_1^\star, \tau_2^\star) \\
p(\boldsymbol{\delta}, \boldsymbol{\beta}|c, \boldsymbol{\eta}, \mathcal{D}) &= N(\boldsymbol{m}^\star, \sigma^2 \boldsymbol{V}^\star).
\end{aligned}
$$

Now we know how to sample $p(c|\boldsymbol{\eta}, \mathcal{D})$, $p(\boldsymbol{\delta}, \boldsymbol{\beta}, \sigma^2|c, \mathcal{D})$ but we also need to sample $\boldsymbol{\eta}$. A good way to do this is by using a Metropolis-Hastings update (Metropolis *et al.* (1953), Hastings (1970)). This cycles through all the elements of $\boldsymbol{\eta}$ in turn, proposing new η_i values from a normal distribution with mean $\boldsymbol{d}'_{\boldsymbol{x}_i}\boldsymbol{\delta} + \boldsymbol{b}'_{\boldsymbol{x}_i}\boldsymbol{\beta}$ and variance σ^2. The proposed value of η_i is then accepted with the minimum of one and the ratio of Poisson likelihoods as the proposal distribution is symmetric. If the residual variation σ^2 turns out to be small then this can affect the updating as the conditional distribution $p(\boldsymbol{\delta}, \boldsymbol{\beta}|c, \boldsymbol{\eta}, \mathcal{D})$ will be highly peaked. In such cases we would recommend updating via an independence-Metropolis algorithm with the proposal distribution taken as the normal approximation to the posterior density. That is, a normal distribution centred at the mode of $p(\boldsymbol{\delta}, \boldsymbol{\beta}|c, \mathcal{D})$ with variance-covariance matrix taken to be the inverse Hessian at this point.

A further simple modification of the model allows us to include or exclude the measured covariates. Dropping a covariate is essentially identical to removing a centre (they both remove a column from $\boldsymbol{B}$) so *BIRTH* and *DEATH* steps for the covariates can be constructed which are very similar to those used for the centres. In addition, this general methodology can also allow us to incorporate nonlinear models for the covariate effects. For instance, if we had just one covariate we could model the effect of this covariate, x_1^c, through a sum of truncated linear spline terms, as we can write

$$
\boldsymbol{b}'_{\boldsymbol{x}}\boldsymbol{\beta} = \sum_{i=1}^{q} \beta_i (x_1^c - t_i)_+, \tag{7.12}
$$

where $(a)_+$ is equal to a when $a > 0$ and zero otherwise and $t_1, \ldots, t_q$ are a set of unknown "knot" points. Here, the effect of the covariate is considered to be a continuous piecewise linear function. Sampling q and the t_i is identical to sampling which covariates to include when we have more than one of them, so this model does not present any new technical difficulties. Further, it could also be used when we have more than one covariate in a Bayesian generalized additive, or even a Bayesian multivariate adaptive regression spline, model (see Denison et al. (2002b) for more details). Nevertheless in the next section, as the dataset we shall analyse has only one covariate, we shall only use the truncated linear spline model in (7.12) to model its effect.

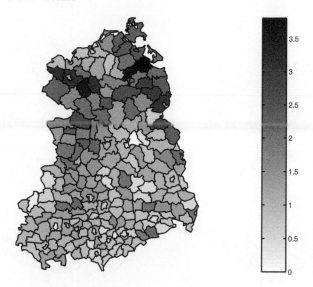

Figure 7.4 *The standardised mortality ratio for the German lip cancer dataset.*

7.4.3 East German Lip Cancer Dataset

We now demonstrate the methodology outlined in the previous section on the German lip cancer dataset studied in Lawson (2001). It concerns the number of deaths due to lip cancer in East Germany during the period 1980–1989. The data consists of the expected number of cases, E_i, and the observed number of cases, y_i, in the 219 spatial regions that make up East Germany. In Fig. 7.4 we plot the standardised mortality ratio (given by y_i/E_i) in each region. Note that the white region halfway up the map and towards its right-hand side is West Berlin, for which no data exist. As is typical with such data we see that there is wide variability between the SMRs and no obvious spatial pattern emerges apart from the fact that rates seem to be lower in the south. The aim of the analysis is to smooth out this raw data to identify a more realistic model for the spatial variation.

As well as the expected and observed counts for the regions the data includes information on a possibly important covariate AFF, the percentage of agricultural, fisheries and forestry workers. This covariate was measured as it is thought that exposure to sunlight, which is an occupational hazard for the workers included, is a risk factor for lip cancer.

In Fig. 7.5 and Fig. 7.6 we display the estimated relative risks in each area found using the Bayesian partition model outlined in Section 7.4.2 when we run the model including, and not including, the information on the covariate AFF. The prior parameters chosen were $\tau_1 = \tau_2 = 10^{-5}$ and $v = 2.5^2$, so that approximately 95% of the prior probability on the levels

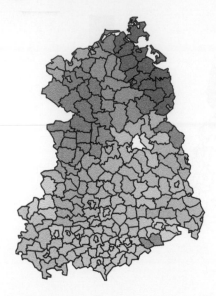 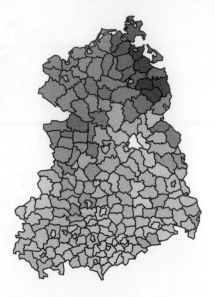

Figure 7.5 *The estimated relative risk in each region for the model when the covariate was not included and using the same color-scale as in Fig. 7.4.*

Figure 7.6 *Same as Fig. 7.5 but with the covariate AFF included in the model.*

changes the relative risk by at most a factor of 5. Note that, to fit easily within the partition model framework, we take each piece of regional data as being given at the centroid of its region. Further, we assumed that the effect of the covariate can be accurately modelled by a truncated linear spline with an unknown number of knots. As might be expected, when we include the covariate we find that the relative risk varies more spatially, due to neighbouring regions perhaps having very different AFF values. This is especially well demonstrated in the north-east region of the map where there is considerable smoothing across the regions in Fig. 7.5, while there are far greater differences between the risks in the same area of Fig. 7.6.

In Fig. 7.7 we display the estimated relationship between the log of the relative risk of lip cancer and the value of AFF. It appears that the covariate has little effect for values below 2 and after that the effect is linear. However, we should not use this figure to interpret that the effect of AFF increases the log relative risk by around 0.18 for values between 0 and 2. This is because the intercept of the graph is only very weakly identifiable by the prior as an increase in it, and a similar decrease in the mean effect of the spatial component, leads to no change in the overall likelihood and probably only a very small change in the prior. Note that the overall effect of the covariate appears smooth, even though each model produces a piecewise

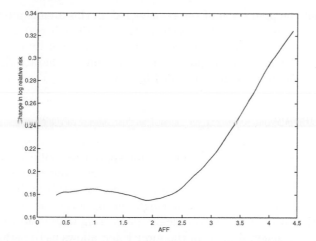

Figure 7.7 *The estimated relationship between the logarithm of the relative risk of disease and the covariate AFF (the % of agricultural, fisheries and forestry workers).*

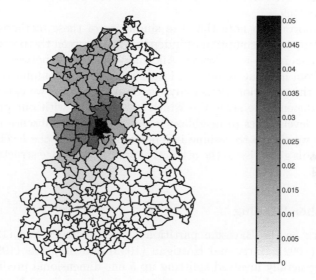

Figure 7.8 *The equivalent kernel found using the partition model without the covariate.*

linear curve. This is due to the posterior averaging over all the models in the generated sample.

Fig. 7.8 gives the equivalent kernel which gives an indication of how important each area was to the darkest one near the centre of the map.

We see that the prediction of the relative risk in this area is predominantly made by those regions in the north-west. Even though regions south of it are nearby in distance terms, their relative risks are hardly used in finding the estimated risk in the region of interest. Similar characteristics of the partition model were seen earlier in Fig. 7.3.

7.5 Discussion

We have summarised some of the properties of Bayesian partition models that make them particularly suitable for spatial modelling. Even though the individual models of the response are discontinuous we have seen how, when the centre locations have been integrated out, this leads to a sensible prior on the dependence between neighbouring locations (i.e. a spherical one with the correlation monotonically decreasing with distance). We have also outlined a generalisation of the model that allows us to perform disease mapping, both with and without covariates. The same methodology also allows us to use nonlinear functions to model the effect of the covariate on the log relative risk of disease. Through the application of the model to two datasets we have seen how it performs for both point-level data and that given in regions.

It is important to note that the simplicity of these methods is highly dependent on the assumption of independence between the process in each of the regions in the tessellation. Allowing dependence between them prevents us from integrating out the levels, significantly complicating simulation from the posterior. Personally, we do not feel that we can accurately specify a dependence structure which is consistent with our prior beliefs about the levels/rates in neighbouring regions. In the absence of such information we choose to assume independence between the levels. We also note that this gives rise to the attractive prior dependence structure as seen in Fig. 7.2.

7.6 Further Reading

Early work on the Bayesian partition models includes Yao (1984), Carlin et al. (1992), Barry and Hartigan (1993) and Stephens (1994). These methods generally involved splitting up a one-dimensional predictor space into independent blocks, equivalently to a one-dimensional Voronoi tessellation. In a spatial context there are numerous examples of partition-like models, e.g. Byers (1992), Ghosh et al. (1997), Sambridge (1999a, 1999b) and Møller and Skare (2001). Other related work includes Silverman and Cooper (1988), Phillips and Smith (1994), Aykroyd (1995), Green (1995), Nicholls (1998) and Rue and Hurn (1999). These papers in image analysis tend to use specifically designed templates to partition the image of interest.

More closely related to the work presented here are the papers on disease mapping given by Knorr-Held and Rasser (2000) and Green and Richardson (2000). Knorr-Held and Rasser (2000) define a partitioning structure on the regions and cluster them into regions of identical risk. See also the follow-up paper by Giudici et al. (2000). Instead, Green and Richardson (2000) model the relative risk using mixtures of Poisson distributions where the probability of being assigned to each mixture depends on their spatial location.

Other papers have utilised Voronoi tessellations to model point process data thought to have been generated by a Poisson rate parameter that varies across the space of interest. Similarly to the partition model introduced here, these assume that the rate is fixed in each region. The papers by Heikkinen and Arjas (1998, 1999) suggested this method and used a specially designed prior to model the relationship between the rates in neighbouring regions. An alternative approach is suggested in Chapter 6 who, similarly to the model outlined here, took the rates in each region to be independent *a priori*.

Acknowledgments

The authors would like to thank Leo Knorr-Held for his comments on an earlier draft of this paper. The first author was supported by grant SFRH BD 1399 2000 from Fundação para a Ciência e Tecnologia, Portugal and the second author acknowledges the support of an EPSRC grant.

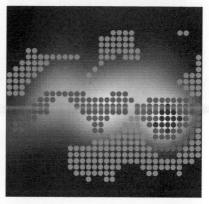

COLOR FIGURE 2.3 S^3 analysis of the earthquake data (shown as a scatterplot in Figure 2.2, see text), using gradient streamlines (left panel) and curvature dots (right panel). The right cluster is clearly statistically significant, the middle cluster is not quite conclusive, and the left cluster is less well defined.

COLOR FIGURE 2.4 S^3 analysis of the asymmetric volcano simulated example. Streamline only (left panel) and streamline and contour (right panel) versions. This shows that the addition of contours enhances the interpretability.

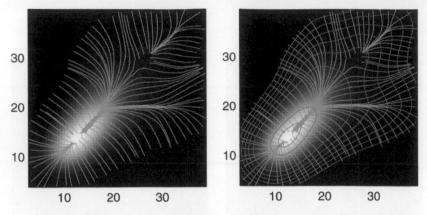

COLOR FIGURE 2.5 Shows streamline and contour analysis is more effective than streamlines only, for the Melbourne temperature data.

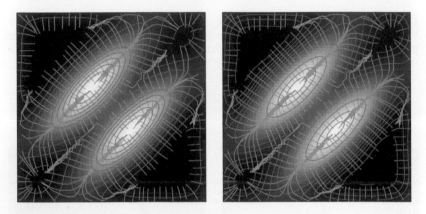

COLOR FIGURE 2.6 Simulated example showing the difference between equal height and modified quantile contour spacing.

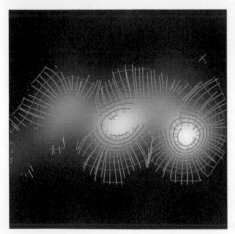

COLOR FIGURE 2.8 Contour and streamline analysis of the earthquake data.

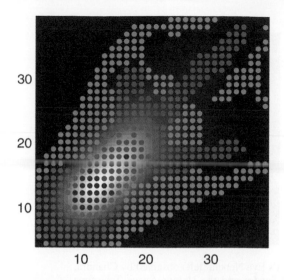

COLOR FIGURE 2.9 Curvature dots analysis of the Melbourne temperature data.

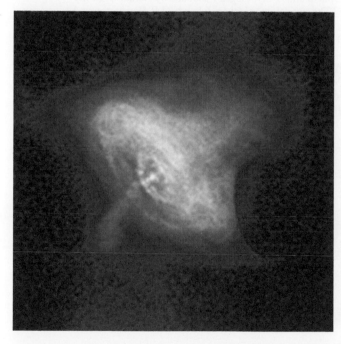

COLOR FIGURE 10.1
This X-ray image of the Crab Nebula is one of the first images sent back by Chandra. The Crab Nebula is one of the youngest and most energetic of about 1000 known pulsars; in fact this supernova remnant produces energy at the rate of 100,000 suns. The image illustrates the extended irregular spatial structure that is typical of Chandra images. (The image was adaptively smoothed; image credit: the National Aeronautics and Space Administration [NASA], the Chandra X-ray Center [CXC], and the Smithsonian Astrophysical Observatory [SAO].)

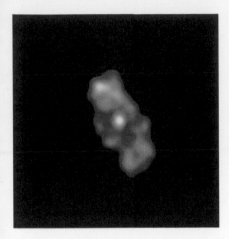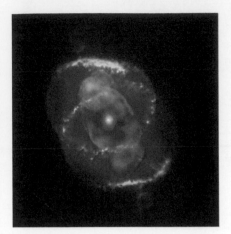

COLOR FIGURE 10.2 X-ray image of the Cat's Eye Nebula (left) observed by Chandra, and a composite X-ray/optical observed by Chandra and the Hubble Space Telescope (HST) (right). The image on the left shows a bright central star, which corresponds to a high density cluster of photons. The composite image on the right illustrates the relative locations of hotter and cooler regions of the planetary nebula. (The X-ray image is adaptively smoothed. The color composite of optical and X-ray images was made by Zoltan G. Levay [Space Telescope Science Institute.] The optical images were taken by J.P. Harrington and K.J. Borkowski [University of Maryland] with HST. Image credits: NASA, University of Illinois at Urbana-Champaign, Chu et al. [2001], and HST.)

COLOR FIGURE 10.3 Centaurus A, a nearby galaxy, as observed by Chandra. The several features of the galaxy — a super-massive black-hole, a jet emanating from the core and the point-like sources scattered about the image — have been clearly resolved for the first time due to the imaging resolution of Chandra and illustrate the variety of photon clusters that might be present in a Chandra image. (The image is adaptively smoothed and exposure corrected; image credit: NASA, SAO, and Kraft et al. [2001].)

CHAPTER 8

Cluster Modelling for Disease Rate Mapping

R.E. Gangnon   M.K. Clayton

8.1 Introduction

Statistical methods for analyzing spatial patterns of disease incidence or mortality have been of great interest over the past decade. To a large extent, the statistical approaches taken fall into two classes: cluster detection or disease mapping. In cluster detection, one typically adopts the hypothesis testing framework, testing the null hypothesis of a common disease rate across the study region against a "clustering" alternative (Whittemore et al. 1987, Kulldorff and Nagarwalla 1995). In disease mapping, one typically uses Bayes or empirical Bayes methods to produce smoothed estimates of the cell-specific disease rates suitable for mapping (see, for example, Clayton and Kaldor (1987) and Besag et al. (1991)). In this paper, we describe a method for inference that simultaneously addresses both the cluster detection and the disease mapping problems.

Many Bayesian approaches to analyzing spatial disease patterns focus on mapping spatially smoothed disease rates (for example, Clayton and Kaldor (1987), Besag et al. (1991) and Waller et al. (1997a)). Mapping methods produce stable estimates for cell-specific rates by borrowing strength from neighboring cells. These are most useful for capturing gradual, regional changes in disease rates, and are less useful in detecting abrupt, localized changes indicative of hot spot clustering. The models proposed by Besag et al. (1991) and Waller et al. (1997a) incorporate both spatially structured (spatial correlation) and unstructured (extra-Poisson variation) heterogeneity in one model. The ability of these models to detect localized clusters is questionable, because they incorporate only a global clustering mechanism. In addition, typically, the spatially structured and unstructured components of the heterogeneity are not separately identified by the likelihood (Waller et al. 1997a).

A few Bayesian approaches more directly address the disease clustering problem, including Lawson (1995), Lawson and Clark (1999b), and Gangnon and Clayton (2000). Lawson (1995) proposes a point process model for detection of cluster locations when exact case (and control) locations are known. Lawson (2000) describes an extension of this model to incorporate both localized clustering and general spatial heterogeneity of disease rates. Lawson and Clark (1999b) describe the application of a point process clustering model to case count data through data augmentation. To apply their model, one imputes locations for each member of the population at risk, typically by assuming a uniform spatial distribution within each cell, to produce a point process. One then proposes a clustering model for the point process. Gangnon and Clayton (2000) propose a model for clustering using cell count data in which the study region is divided into several components: a large background area and a relatively small number of clusters where a common rate (or covariate-adjusted risk) is assumed within each component.

Knorr-Held and Rasser (2000) and Denison and Holmes (2001) consider a nonparametric Bayesian framework for modelling cell count data. Although superficially similar to the Gangnon and Clayton (2001) model in that cells are grouped into components of constant risk, the models of Knorr-Held and Rasser (2000) and Denison and Holmes (2001) serve a very different goal. In the models of Knorr-Held and Rasser (2000) and Denison and Holmes (2001), the components (or clusters) of cells primarily serve as a tool for estimating the underlying risk surface, not as parameters of direct interest. In the model of Gangnon and Clayton (2000), the location and composition of the cluster of cells is of primary interest.

In Section 8.2, we describe the model for clustering originally proposed by Gangnon and Clayton (2000). In Section 8.3, we review a randomized variant on a backwards selection algorithm useful for approximating the posterior distribution on cluster models. In Section 8.4, we illustrate the use of these modelling techniques through the construction of maps of the 1995 mortality rates from five cancers (breast, cervical, colon, lung and stomach) in the United States.

8.2 Statistical Model

Consider a study region divided into N subregions, or cells. A cell is typically a geopolitical subregion of the study region, e.g., a state within a country, a county within a state, a census tract or block within a county. For each cell i, we observe y_i, the number of cases of disease (or deaths), and E_i, the expected number of cases (or deaths) calculated using either internal or external standardization. We assume a Poisson model for the data, i.e., $y_i \sim \text{Poisson}(r_i E_i)$, where r_i is the standardized incidence (or mortality) ratio for cell i.

To draw inferences about r_i, we propose a Bayesian model with a hierarchical prior. First, divide the cells into $k + 1$ groups, or components. Call one of these components the background and the other k components clusters. Denote the set of cells belonging to the background by $\mathbf{c}_0$, and the sets of cells belonging to the clusters by $\mathbf{c}_1, \mathbf{c}_2, \ldots, \mathbf{c}_k$. The $k!$ models with the same cluster components are regarded as identical, and a unique representation is chosen (for example, we typically require that $\min \mathbf{c}_1 < \min \mathbf{c}_2 < \ldots < \min \mathbf{c}_k$). Given a specific cluster model $\mathbf{c}_0, \mathbf{c}_1, \ldots, \mathbf{c}_k$, we assume that cells belonging to component j share a common risk ratio ρ_j, i.e., $r_i = \sum_{j=0}^{k} \rho_j I_{\{i \in \mathbf{c}_j\}}$.

To construct our model, we temporarily assume that $\mathbf{c}$ is known. Given the arbitrary labeling of the clusters, it is sensible to utilize an exchangeable prior for $\rho_1, \rho_2, \ldots, \rho_k$. Specifically, in our examples, we take $\rho_0 \mid \mathbf{c} \sim Gamma(\alpha_0, \beta_0)$, and $\rho_j \mid \mathbf{c} \sim Gamma(\alpha, \beta), j = 1, 2, \ldots, k$.

Given $\mathbf{y}$ and $\mathbf{c}$, $\rho_0, \rho_1, \ldots, \rho_k$ are independent, $\rho_0 \mid \mathbf{c}, \mathbf{y} \sim Gamma(\alpha_0 + y_{\bullet 0}, \beta_0 + E_{\bullet 0})$, and $\rho_j \mid \mathbf{c}, \mathbf{y} \sim Gamma(\alpha + y_{\bullet j}, \beta + E_{\bullet j}), j = 1, 2, \ldots, k$, where $y_{\bullet j} = \sum_{i=1}^{N} y_i \cdot I_{\{c_i = j\}}$ is the total number of cases in cluster j, $E_{\bullet j} = \sum_{i=1}^{N} E_i \cdot I_{\{c_i = j\}}$ is the total expected number of cases in cluster j, and $I_{\{c_i = j\}}$ is 1 if $c_i = j$ and 0 if $c_i \neq j$. The marginal likelihood of $\mathbf{y}$ given $\mathbf{c}$ is

$$p(\mathbf{y}|\mathbf{c}) = \frac{\Gamma(\alpha_0 + y_{\bullet 0})}{\Gamma(\alpha_0)} \cdot \frac{\beta_0^{\alpha_0}}{(\beta_0 + E_{\bullet 0})^{\alpha_0 + y_{\bullet 0}}} \cdot$$

$$\prod_{j=1}^{k} \frac{\Gamma(\alpha + y_{\bullet j})}{\Gamma(\alpha)} \cdot \frac{\beta^{\alpha}}{(\beta + E_{\bullet j})^{\alpha + y_{\bullet j}}} \cdot \prod_{i=1}^{N} \frac{E_i^{y_i}}{y_i!}. \tag{8.1}$$

This closed-form expression for the marginal likelihood allows us to avoid explicit consideration of the risk parameters in calculating the posterior distribution over cluster models.

Next, given k (i.e., the number of clusters), we define a prior distribution for the cluster model $\mathbf{c}_0, \mathbf{c}_1, \ldots, \mathbf{c}_k$ using a product specification. That is, the prior probability of a particular cluster model is given by

$$p(\mathbf{c}_0, \mathbf{c}_1, \ldots, \mathbf{c}_k | k) = \exp\left(-\sum_{j=1}^{k} S(\mathbf{c}_j)\right),$$

where S is a function for scoring the prior probability of a cluster based on its geographic properties. The particular scoring function should be tailored to the specific application. For example, in an analysis of leukemia incidence in census tracts or blocks within an eight-county region of New York state, Gangnon and Clayton (2000) develop a scoring function based on area and perimeter of clusters designed to detect small, circular clusters. For the cancer mortality atlas example in Section 8.4, we consider an alternative specification of the prior based on a simpler measure of cluster size.

Our measure of cluster size requires only knowledge of the adjacency structure amongst the cells. Knowledge of the adjacency structure amongst the cells allows us to think of the map as an undirected graph. The distance between two cells in this graph is defined to be the length of the shortest path between the two cells in the graph. For any cluster (i.e., connected subgraph), we define the cluster size to be the maximum distance in the entire graph between any pair of cells belonging to the cluster; cluster size ranges between 0 and s_{max}, where s_{max} is the diameter of the graph, i.e., the maximum distance between any pair of cells. (An alternative measure of cluster size would be the diameter of the induced connected subgraph generated by the cluster so that distances between cells are measured based only on paths entirely contained within the cluster. Our definition of cluster size has some operational advantages, and the differences between the two definitions are generally quite small.)

We now discuss one possible scoring function for cluster size, which is motivated by a scheme for randomly selecting a cluster. For this scheme, we first select a cluster size and then, conditional on the chosen cluster size, select a cluster at random. First, the cluster size is chosen at random, with probability p_s, $s = 0, 2, \ldots, s_{max}$. The cluster is then selected at random, with probability $1/n_s$ from the n_s clusters of the specified size. Thus, the probability of selecting a cluster $\mathbf{c}$ of size s under this scheme is $1/(p_s n_s)$, and the corresponding scoring function is $S(\mathbf{c}) = -\log(p_s) - \log(n_s)$. Calculation of the n_s can be computationally burdensome for even modest values of s because n_s typically increases exponentially with s. For the example in Section 8.4, we explicitly calculate n_0 and n_1 and use a rough extrapolation to approximate n_s for larger values of s.

8.3 Posterior Calculation

Given the large number of potential cluster models, we cannot, in practice, evaluate the desired posterior through direct enumeration. Instead, we propose using a randomized variant of an agglomerative clustering algorithm (or a backwards variable selection algorithm) to calculate an approximation to the posterior. The first component of this algorithm is Occam's window (Madigan and Raftery 1994, Raftery et al. 1997), which serves to reduce consideration to a relatively small number of models over which a model average can be computed.

In the symmetric version of Occam's window, one excludes a model M from consideration if there exists another model M' such that

$$\frac{p(M|\text{data})}{p(M'|\text{data})} < 1/W$$

for a fixed W. In the strict version of Occam's window, one also excludes a model M from consideration if there exists a sub-model M_s of M with $p(M_s|\text{data}) \geq p(M|\text{data})$. This second rule is an attempt at an implementation of Occam's razor, or the principle of parsimony.

In most applications, we utilize the symmetric version of Occam's window. If the chosen W is sufficiently large, the symmetric version of Occam's window only serves to truncate the far tails of the posterior and thus should not have an excessive impact on our inferences. On the other hand, the strict rule based on Occam's razor, while intuitively appealing, potentially excludes from consideration very competitive models and thus can have a dramatic effect on our inferences. For example, let two models M_0 and M_1 have posterior probabilities of 0.51 and 0.49, respectively. If M_0 is a sub-model of M_1 and we use the strict version of Occam's window, the "posterior" probability of M_0 becomes 1 and our large degree of uncertainty about the correct model is translated into certainty that the simpler model is correct.

To find the models falling within the symmetric version of Occam's window (with parameter W), we utilize a randomized search algorithm similar to both backwards elimination methods for variable selection or agglomerative methods for clustering. In our algorithm, we start with one of the N saturated models, i.e., one of the models with $N - 1$ clusters, and repeatedly merge adjacent components of the current model with the aim of producing models with high posterior probability.

Given the size of the model space, we will not be able to determine with certainty which mergers lead to better models. To address this, let $\mathbf{c}_0, \mathbf{c}_1, \ldots, \mathbf{c}_k$ be the components of the current model, and let p_{ij} be the posterior probability of the model obtained by merging $\mathbf{c}_i$ and $\mathbf{c}_j$, $i = 0, 1, \ldots, k-1$, $j = i+1, i+2, \ldots, k$. Note that $p_{ij} \equiv 0$ for the components $\mathbf{c}_i$ and $\mathbf{c}_j$ that are not adjacent to each other. The fundamental notion behind our search algorithm is that the posterior density p_{ij} is a good proxy measure for the likelihood that merging $\mathbf{c}_i$ and $\mathbf{c}_j$ will eventually lead to models with high posterior probabilities. To justify this notion, note that the posterior probability p_{ij} can be thought of as a measure of the relative probability that the disease rates for components $\mathbf{c}_i$ and $\mathbf{c}_j$ are the same, or equivalently that the cells in components $\mathbf{c}_i$ and $\mathbf{c}_j$ belong to the same component in the true cluster model.

In practice, we select a merger of adjacent components using the following three-step procedure.

1. For each pair of adjacent components $\mathbf{c}_i$ and $\mathbf{c}_j$, $i < j$, in the current model M, we calculate

$$
\begin{aligned}
\mathcal{M}_{ij} &= \frac{p_{ij}}{p(M)} \\
&= \frac{\Gamma(\alpha_i + y_{\bullet i} + y_{\bullet j})\Gamma(\alpha_j)}{\Gamma(\alpha_i + y_{\bullet i})\Gamma(\alpha + y_{\bullet j})} \frac{(\beta_i + E_{\bullet i})^{\alpha_i + y_{\bullet i}}(\beta + E_{\bullet j})^{\alpha + y_{\bullet j}}}{(\beta_i + E_{\bullet i} + E_{\bullet j})^{\alpha_i + y_{\bullet i} + y_{\bullet j}}\beta^\alpha},
\end{aligned}
$$

where $p(M)$ is the posterior probability of the current model M and $(\alpha_i, \beta_i) = (\alpha, \beta)$ for $i > 0$. We note that $\mathcal{M}_{ij}$ depends the current model only through the components c_i and c_j. Thus, after each merger, we need only update the M_{ij} if component c_i or c_j was involved in the merger.

2. Calculate the probability of merging components c_i and c_j, $i < j$, denoted by $\mathcal{P}_{ij}$, by truncating $\mathcal{M}_{ij}$ using a symmetric Occam's window with parameter W' (in most applications, we take $W' \leq W$), i.e., calculate and then normalize the values

$$\mathcal{P}_{ij} \propto \mathcal{M}_{ij} \quad \text{if} \quad \frac{\mathcal{M}_{ij}}{\max_{(i,j)} \mathcal{M}_{ij}} \geq 1/W'$$

$$= 0 \quad \text{otherwise}$$

Truncating the merger probabilities using Occam's window helps prevent a large number of poor mergers from overwhelming a small number of promising mergers, and can, in practice, greatly speed the model search.

3. Select the merger of adjacent components c_i and c_j, $i < j$, with probability $\mathcal{P}_{ij}$.

Our search algorithm proceeds by repeatedly applying the merger selection step discussed above. To speed the search, we first reduce its scope. Instead of beginning each new search from a saturated model, we first build a smaller, but still implausibly large, cluster model as a base (model) for multiple searches. By starting searches from smaller base models, we increase the number of good models that can be examined in a fixed time period. Once searches from a given base stop producing new, good models, we build a new base from which the search can proceed. We outline the details of the search algorithm with a single base (size). The extension to multiple bases is straightforward.

1. Start with one of the N saturated models which contain $N - 1$ clusters and a single background cell by selecting one of the N cells at random to be the background cell. Denote this model by $M_s = \{c_0^s, c_1^s, \ldots, c_{N-1}^s\}$.

2. Repeatedly merge adjacent components of M_s using the merger selection procedure described above to produce a model with k clusters. Denote this model by $M_b = \{c_0^b, c_1^b, \ldots, c_k^b\}$.

3. Repeatedly merge adjacent components of M_b using the merger selection procedure described above to produce a nested series of cluster models with $k - 1, k - 2, \ldots, 2, 1$ clusters.

4. Update the current list of cluster models falling within the symmetric version of Occam's window (with parameter W) to reflect the models found in Step 3.

5. Repeat Steps 3 and 4 until a prespecified stopping rule is satisfied.

6. Repeat Steps 1-5 until a prespecified stopping rule is satisfied.

Possible stopping rules for Steps 5 and 6 include stopping after a fixed number of iterations or stopping after some number of consecutive failures to find a new model falling within Occam's window. We generally use an adaptive stopping rule based on sequential determination of whether the success probability θ in a sequence of Bernoulli trials is zero. With a non-trivial prior and loss function, the optimal Bayes sequential rule is to stop after observing the first success or when the posterior mean for θ is less than the cost of the next observation (Gangnon 1998). In the former case, we will begin a new sequence of searches with a different θ; in the latter case, we will stop the search. With good prior choices, we can quickly abandon poor bases and exhaustively sample good bases.

Using the approximate posterior distribution based on the models falling within Occam's window, we can estimate various cell-specific quantities. For example, a reasonable estimate for the cell-specific standardized incidence (or mortality) ratio $\mathbf{r}$ is its posterior mean

$$\hat{r}_i = \mathrm{E}(r_i|\mathbf{y}) = \sum_{\mathbf{c}} \mathrm{E}(\rho_{c_i}|\mathbf{y}, \mathbf{c}) \cdot p(\mathbf{c}|\mathbf{y}), i = 1, 2, \ldots, N.$$

In contrast to estimates based on a single cluster model, the posterior means will generally produce "fuzzy" edges of clusters, reflecting uncertainty about the exact composition of the clusters. A grayscale map of the $\hat{r}_i$'s can effectively demonstrate both cluster locations and cluster risks. In cases where the cluster borders are of primary interest, this map may be supplemented by a map of the posterior probability that the cell belongs to a cluster,

$$p(c_i > 0|\mathbf{y}) = \sum_{\mathbf{c}:c_i>0} p(\mathbf{c}|\mathbf{y}), i = 1, 2, \ldots, N.$$

8.4 Example: U.S. Cancer Mortality Atlas

The Centers for Disease Control and Prevention (CDC) provide access to the Compressed Mortality File (CMF), a county-level national mortality and population data base spanning the years 1979-98, through the CDC WONDER website (http://wonder.cdc.gov). From this data base, we obtained the number of adult ($\geq$ 20 years old) deaths in 1995 from five cancers for the 48 contiguous states and the District of Columbia (DC). The five cancer sites were lung, female breast, cervical, colorectal and stomach. For each state, an expected death count from each cancer was calculated by applying the 1995 national death rates from each cancer for subgroups based on age, sex (where appropriate) and race to the state's population.

In this series of analyses, we wish to identify regions of clustering of the state-specific mortality rates within the maps. To guide the analysis, we use a particular implementation of the scoring function discussed in Section 8.2. The maximum distance between any two states measured within the corresponding graph is 11. So, the potential cluster sizes range between 0 and 11. We select the cluster size uniformly from the available cluster sizes, i.e., $p_s = 1/11$, $s = 0, 1, \ldots, 11$. As noted previously, explicit calculation of the number of clusters is difficult because the number of clusters grows exponentially along with the cluster size. Instead, we calculate only n_0 and n_1 explicitly. For larger values of s, we use the value $n_s = (s-1)!c^s n_0$, where $c = n_1/n_0$. (This formula is the solution of the recurrence relation $n_s = (s-1)cn_{s-1}$. In this formula, $(s-1)c$ represents the multiplicative increase in the number of clusters as cluster size increases. The factor $s-1$ represents a crude approximation to the number of cells in the clusters so that c is the "per cell" multiplier.) For the U.S. map, $n_0 = 49$, $n_1 = 162$ and $c = 3.31$.

For these analyses, we make no prior assumptions regarding the direction of risks within clusters, and so use the same gamma prior for both the background risk and the cluster risk. A priori, we have no reason to believe that the expected number of deaths given for each state is incorrect. This belief leads us to specify a $Gamma(\gamma, \gamma)$ prior for the risks for some $\gamma > 0$. Given the relatively small number of clusters in most models, the data will not provide enough information to reliably estimate γ. To evaluate the impact of the choice of γ on our inference, we analyzed each data set using several different values for γ $(0.1, 0.5, 1, 2, 10)$. The choice of γ had little impact on our inferences, and thus results are presented for $\gamma = 1$.

For the posterior approximations, we use a symmetric Occam's window with parameter $W = 1000$. To speed the search, we utilize two bases (or base sizes), 48 clusters and 26 clusters. Thus, we implicitly assumed the maximum number of clusters in a model belonging to Occam's window was 25. In practice, the number of clusters never approached this limit. If it had, we would have restarted the search with a larger second base.

The estimated cancer mortality risks by state are displayed in Figure 8.1 for (female) breast cancer, in Figure 8.2 for cervical cancer, in Figure 8.3 for colorectal cancer, Figure 8.4 for lung cancer and in Figure 8.5 for stomach cancer. Each figure includes a grayscale map of the contiguous United States. The grayscale indicates the estimated, or posterior mean, risk for each state. (The grayscale is consistent across the figures.) Below each map, we provide equal-tailed 95% posterior intervals for the mortality risk in each state based on the model-averaged posterior distributions for the state-specific mortality risks. (In these displays, the states are ordered by the posterior mean risk.)

8.4.1 Breast Cancer

Examining Figure 8.1, we observe that states appear to fall into one of three groups – a higher risk group of states in the Northeast and Midwest with posterior mean mortality ratios of approximately 1.06, a lower risk group of states in the South and Midwest with posterior mean mortality ratios of approximately 0.95, and a group of border states which could potentially belong to either group. The border states could be further subdivided into the states more likely to belong to the higher risk group (e.g., Missouri, North Dakota, Montana, D.C. and Indiana) and states more likely to belong to the lower risk group (e.g., Oklahoma, South Dakota, West Virginia, Kansas, Idaho, Virginia, Nebraska, Washington). For states in the third group, the uncertainty about cluster membership produces a bimodal posterior for the mortality ratio, which translates into very wide posterior intervals for the state-specific mortality risks.

8.4.2 Cervical Cancer

In Figure 8.2, we observe clear evidence of a single cluster of states with elevated risk of cervical cancer mortality. Five states clearly belong to this cluster: Arkansas, Kentucky, West Virginia, Tennessee and Texas. The (posterior mean) increase in mortality associated with this cluster is approximately 27%. Up to four other states – Indiana, Louisiana, Oklahoma and New Mexico – are likely to also belong to this cluster, and up to five additional states beyond those four could possibly belong to the cluster.

8.4.3 Colorectal Cancer

For colorectal cancer (Figure 8.3), the states fall into three groups based on the mortality risk levels. Note, however, that the majority of models included in Occam's window contain four clusters indicating that some of the similar risk groups are disconnected. The higher risk group, with a posterior mean risk of approximately 1.12, consists of 14 states in the Northeast and Midwest. The lower risk group, with an approximate mortality risk of 0.90, consists of a cluster of 3 states in the Southeast (Alabama, Florida, Georgia) and a cluster of 11 states in the West. The third medium risk group, with an approximate colorectal cancer mortality risk of 1.00, consists of 8 states in the Midwest and South. The majority of the other states fall on the border between two of these three groups and could conceivably belong to either group. Two states are clear exceptions to this rule – Louisiana and Connecticut. Louisiana is an isolated higher risk state within the two lower risk groups; Connecticut is an isolated lower risk state within the Northeast. The wide posterior intervals for these two states primarily reflect the uncertainty associated with estimates for state-specific risks, when we are unable to borrow strength from neighboring states

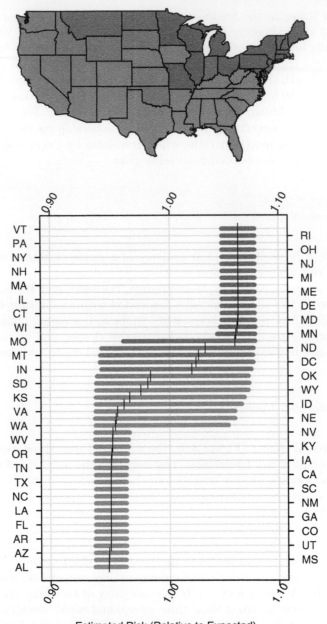

Figure 8.1 *Estimated female breast cancer mortality risk (relative to expected number of deaths based on age-race distribution) by state (1995). Grayscale indicates posterior mean. Intervals are equal-tailed 95% posterior probability intervals. Vertical lines indicate the posterior mean.*

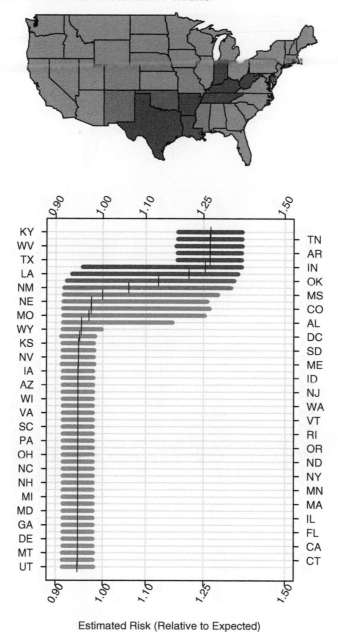

Figure 8.2 *Estimated cervical cancer mortality risk (relative to expected number of deaths based on age-race distribution) by state (1995). Grayscale indicates posterior mean. Intervals are equal-tailed 95% posterior probability intervals. Vertical lines indicate the posterior mean.*

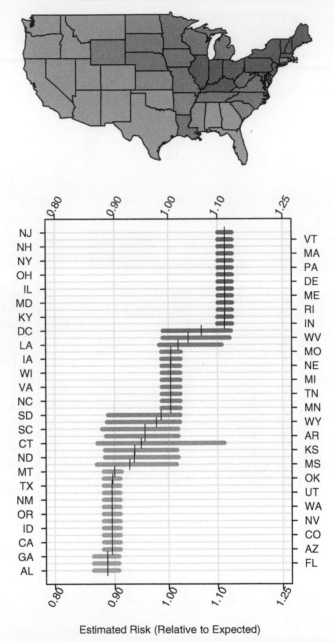

Figure 8.3 *Estimated colorectal cancer mortality risk (relative to expected number of deaths based on age-sex-race distribution) by state (1995). Grayscale indicates posterior mean. Intervals are equal-tailed 95% posterior probability intervals. Vertical lines indicate the posterior mean.*

8.4.4 Lung Cancer

For lung cancer (Figure 8.4), we observe wide variations in mortality risks across the entire United States. Models for lung cancer risks falling within Occam's window contain between 12 and 18 clusters. The lowest lung cancer mortality risk, by far, occurs in Utah (0.42); Colorado and New Mexico also have relatively low risks. Lower than expected mortality risks are also evident across much of the North. The highest lung cancer mortality risks occur in Kentucky, Delaware and West Virginia. Other areas with higher than expected risks are a cluster consisting of Arkansas, Louisiana, Tennessee, Missouri and Oklahoma and the state of Maine. Presumably, the large variations in lung cancer mortality risks across the map are due to similar variations in smoking rates across the country, but we do not currently have the data needed to verify this conjecture.

8.4.5 Stomach Cancer

For stomach cancer (Figure 8.5), we observe three prominent clusters, two with higher than expected mortality and one with lower. The first higher risk cluster consists of eight states in the Northeast (three additional states – Delaware, Maryland and the District of Columbia could belong to this cluster). The second higher risk cluster is California (possibly joined by Nevada). The lower risk cluster consists of 4–10 states in the center of the country.

8.5 Conclusions

In this paper, we presented a model for spatial clustering of disease in which the region of space under study is partitioned into a single large background area and a relatively small number of clusters. We adopted a Bayesian framework for inference, using a simple, but approximate, prior specification of the probabilities of various clusters to guide, but not necessarily limit, selection from the available models. We utilized a randomized version of an agglomerative clustering algorithm to find models falling within the symmetric version of Occam's window.

In Section 8.4, we illustrated the use of this modelling approach by constructing maps of the 1995 mortality rates from five cancers in the United States. The examples in this section show that, with our cluster modelling approach, we can produce a map with many clusters, e.g., the map for lung cancer mortality (Figure 8.4), a map with few clusters, e.g., the map for stomach cancer mortality (Figure 8.5), and a map with only a single cluster, e.g., the map for breast cancer (Figure 8.1). The variety of different map appearances tends to indicate that our estimates are an honest reflection of the underlying data, not solely a reflection of our prior assumptions. This

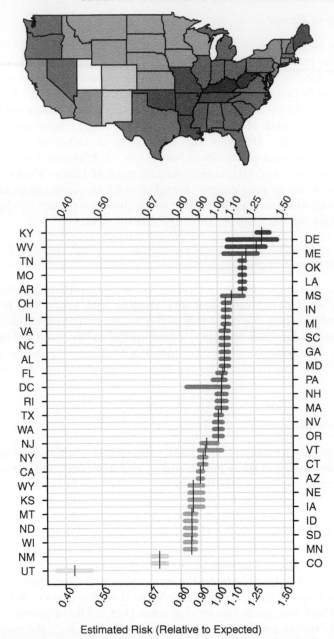

Figure 8.4 *Estimated lung cancer mortality risk (relative to expected number of deaths based on age-sex-race distribution) by state (1995). Grayscale indicates posterior mean. Intervals are equal-tailed 95% posterior probability intervals. Vertical lines indicate the posterior mean.*

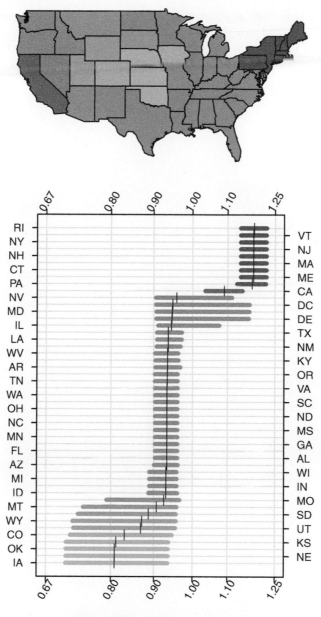

Figure 8.5 *Estimated stomach cancer mortality risk (relative to expected number of deaths based on age-sex-race distribution) by state (1995). Grayscale indicates posterior mean. Intervals are equal-tailed 95% posterior probability intervals. Vertical lines indicate the posterior mean.*

contention is further supported by additional analyses of these data that we have performed using either different approximations for n_s or different measures of cluster size (Gangnon 1998). In all cases, although the results differed in some details, all of the major inferences were unaffected by the chosen prior.

CHAPTER 9

Analyzing Spatial Data Using Skew-Gaussian Processes

H. Kim   B.K. Mallick

9.1 Introduction

Spatial statistics has become a tool-box of methods useful for attacking a range of problems in scientific fields such as petroleum engineering, civil engineering, geography, geology, hydrology. The most useful technique for statistical spatial prediction is kriging (Cressie 1993). Most theories related to spatial prediction assume that the data are generated from a Gaussian random field. However non-Gaussian characteristics, such as (non-negative) continuous variables with skewed distribution, appear in many datasets from scientific fields. For example potential, porosity and permeability measurements from petroleum engineering applications usually follow skewed distribution. A common way to model this type of data is to assume that the random field of interest is the result of an unknown nonlinear transformation of a Gaussian random field. Trans-Gaussian kriging is the kriging variant used for prediction in transformed Gaussian random fields, where the normalizing transformation is assumed known. This approach has some potential weaknesses (De Oliveira et al. 1997, Azzalini and Capitanio 1999) such as:

(i) the transformations are usually on each component separately, and achievement of joint normality is only hoped for;

(ii) the transformed variables are more difficult to interpret, especially when each variable is transformed by using a different function;

(iii) even though the normalizing transformation can be estimated by maximum likelihood, it may be unwise to select a particular transformation;

(iv) sometimes the back-transformed fitted model is severely biased (Miller 1984, Cressie 1993).

Alternatively, we can use more general, flexible parametric classes of multivariate distributions to represent features of the dataset aiming to reduce the unrealistic assumptions. The pioneering work in this field started with Zellner (1976) who dealt with the regression model with multivariate Student-t error terms.

To handle skewed data lognormal kriging may be used, but it can only be applied to data which follow a normal distribution after log transformation. Hence, a more general distributional form is needed to deal with real data. Azzalini and Dalla Valle (1996) and Azzalini and Capitanio (1999) developed the skew-Gaussian distribution, which has many similar properties as normal distribution.

In this chapter we suggest using skew Gaussian processes for prediction and illustrate the model on real data, obtaining satisfactory results. To accurately identify clusters, or peaks, in the true underlying spatial process we need to be able to obtain good approximations of it. The accuracy of any model necessarily depends on the assumptions made by it. Here, as the responses of many spatial processes are always non-negative, we present a method for analysing such data that generalizes lognormal kriging.

The chapter is organized as follows. In Section 9.2 we use a skew-Gaussian process to model a somewhat-skewed dataset. We employ a Markov chain Monte Carlo (MCMC) algorithm, an accept-reject algorithm with a latent variable within the Gibbs sampler to generate samples outlined in Section 9.2.3. In Section 9.3, we apply this model to the spatial prediction of flow potential data for the North Burbank Field. We know that this dataset was generated by a skewed distribution after exploratory data analysis. Finally we offer a brief discussion in Section 9.4.

9.2 Skew-Gaussian Processes

If we know that a dataset is generated from a somewhat-skewed distribution then we can do an analysis directly using skewed distributions. One such distribution is the skew-Gaussian one which was proposed by Azzalini and Dalla Valle (1996). Another flexible family is the lognormal distribution which only covers $\lambda = 0$ of a Box-Cox transformation (Box and Cox 1969). Box-Cox transformation is frequently used for "normalizing" positive data:

$$g_\lambda(x) = \begin{cases} \frac{x^\lambda - 1}{\lambda} & \text{if } \lambda \neq 0 \\ log(x) & \text{if } \lambda = 0 \end{cases}.$$

That is the log transformed data follows a normal distribution. The following skew-Gaussian distribution can explain a wider class of skewed data.

The model we shall propose is a special case of Azzalini and Capitanio (1999), and it is proved that any linear combination of skew-Gaussian random variates is still skew-Gaussian. Hence it is natural to define the skew-Gaussian process based on the definition of the Gaussian process (Ripley 1981). A skew-Gaussian process (SGP) is a collection of random variables Y_x indexed by a set $x \in \mathcal{X}$, where any finite collection of Ys has a joint multivariate skew-Gaussian distribution. That is, every linear combination has a skew-Gaussian distribution.

9.2.1 The Model

Azzalini and Capitanio (1999) proposed the multivariate skew-Gaussian distribution. For simplicity, and to avoid the introduction of more parameters, we consider the following model. Let

$$\boldsymbol{Z}^* = (Z(\boldsymbol{x}_0), \boldsymbol{Z}')' \sim SN_{n+1}(F^*\boldsymbol{\beta}, \sigma^2 K_\theta^*, \frac{\alpha}{\sigma}\boldsymbol{1}_{n+1})$$

have density

$$f(\boldsymbol{z}^*|\boldsymbol{\eta}) = 2\phi_{n+1}(\boldsymbol{z}^* - F^*\boldsymbol{\beta}; \sigma^2 K_\theta^*)\Phi\left(\frac{\alpha}{\sigma}\boldsymbol{1}'_{n+1}(\boldsymbol{z}^* - F^*\boldsymbol{\beta})\right), \qquad (9.1)$$

for $\boldsymbol{z}^* \in \mathbb{R}^{n+1}$ and where $\alpha \in \mathbb{R}$ is known as the "shape" parameter (although the actual skewness is regulated in a more complex way) with $\sigma \in \mathbb{R}^+$ a scale parameter. Further, $\boldsymbol{\eta} = (\boldsymbol{\beta}', \sigma^2, \boldsymbol{\theta}', \alpha)'$ and denotes the set of model parameters and $\boldsymbol{1}_k$ is the $k \times k$ matrix with every entry one. Note that when α equals 0, (9.1) reduces to the $N_{n+1}(F^*\boldsymbol{\beta}, \sigma^2 K_\theta^*)$ density.

In general, we observe $\boldsymbol{Z} = (Z(\boldsymbol{x}_1), \cdots, Z(\boldsymbol{x}_n))'$ and we shall focus on the prediction of $Z(\boldsymbol{x}_0)$, where $\boldsymbol{x}_0$ is an unknown spatial location. So the distributional assumption of $\boldsymbol{Z}^*$ is natural, where

$$F^* = \begin{pmatrix} f'(\mathbf{x}_0) \\ F \end{pmatrix}, \qquad K_\theta^* = \begin{pmatrix} k_\theta(0) & k'_\theta \\ k_\theta & K_\theta \end{pmatrix}, \qquad (9.2)$$

$k_\theta(0) = K_\theta(\boldsymbol{x}_0, \boldsymbol{x}_0)$, $k_\theta = [K_\theta(\boldsymbol{x}_0, \boldsymbol{x}_i)]_{n \times 1}$, $F = [f_j(\boldsymbol{x}_i)]_{n \times q}$ and K_θ is an $n \times n$ correlation matrix. Then $\boldsymbol{Z} = (Z(\boldsymbol{x}_1), Z(\boldsymbol{x}_2), \cdots, Z(\boldsymbol{x}_n))'$ is a realization from the skew Gaussian random field on $\mathbb{R}^n$ with similar parameters to those for $\boldsymbol{Z}^*$ but with F^* replaced by F, etc. Hence $[f(\boldsymbol{x})]'\boldsymbol{\beta} = (f_1(\boldsymbol{x}), \cdots, f_q(\boldsymbol{x}))'\boldsymbol{\beta}$. Furthermore $\sigma^2 K_\theta$ is positive definite matrix and $\sigma^2 K_\theta(\boldsymbol{x}, \boldsymbol{y}) = \sigma^2 K_\theta(\|\boldsymbol{x} - \boldsymbol{y}\|)$, where $\boldsymbol{\theta} \in \Theta$ is a $p \times 1$ vector of structural parameters, Θ is a subset of $\mathbb{R}^p$, $\sigma^2 \in \mathbb{R}^+$ and $\|\cdot\|$ denotes Euclidean distance. Structural parameters control the range of correlation and/or the smoothness of the random field, where for every $\boldsymbol{\theta} \in \Theta$, $K_\theta(\cdot)$ is an isotropic correlation function, assumed to be continuous on $\boldsymbol{\theta}$.

As our working family of isotropic correlation functions, we use the general exponential correlation function (Yaglom 1987, De Oliveira et al. 1997).

$$K_\theta(l) = \exp\left(-\nu l^{\theta_2}\right) = \theta_1^{l^{\theta_2}}, \qquad (9.3)$$

where l represents Euclidean distance between two locations, $\nu > 0, \theta_1 = e^{-\nu} \in (0, 1)$, and $\theta_2 \in (0, 2]$ are unknown parameters.

Using Proposition 4 of Azzalini and Dalla Valle (1996), the marginal distribution of $\boldsymbol{z}^*$ turns out to be a skew-Gaussian distribution and the conditional distribution becomes a member of an extended skew-Gaussian class of densities. These are given by the following expressions:

$$f(z|\eta) \;=\; 2\phi_n\left(z - F\beta, \sigma^2 K_\theta\right)\Phi\left(\varphi\right), \qquad z \in \mathbb{R}^n, \qquad (9.4)$$

$$f(z(x_0)|z,\eta) \;=\; \phi_1\left(z(x_0) - \mu_{1.2}, \sigma^2 K_{1.2}\right)$$
$$\times \; \frac{\Phi\left(\frac{\alpha}{\sigma}(z(x_0) - \mu_{1.2}) + \varphi\sqrt{1 + \alpha^2 K_{1.2}}\right)}{\Phi\left(\varphi\right)}, \qquad (9.5)$$

where $z(x_0) \in \mathbb{R}$, $\varphi = \frac{\alpha}{\sigma}\alpha'_z(z - F\beta)$, $\alpha'_z = 1'_{n+1}\left(K_\theta^{-1} k_\theta \;\; I_n\right)' / \sqrt{1 + \alpha^2 K_{1.2}}$, $\mu_{1.2} = f'(x_0)\beta + k'_\theta K_\theta^{-1}(z - F\beta)$, $K_{1.2} = k_\theta(0) - k'_\theta K_\theta^{-1} k_\theta$ and I_n denotes the n-dimensional identity matrix.

9.2.2 Bayesian Analysis

The Bayesian model specification requires prior distributions for all the unknown parameters. We shall assume prior independence so that the joint prior distribution

$$\pi(\eta) = \pi(\beta, \sigma^2, \theta, \alpha) = \pi(\beta)\pi(\sigma^2)\pi(\theta)\pi(\alpha),$$

so that the posterior distribution is proportional to

$$f(z|\eta)\pi(\eta) = f(z|\beta, \sigma^2, \theta, \alpha)\pi(\beta)\pi(\sigma^2)\pi(\theta)\pi(\alpha).$$

Let the prior distribution on β and σ^2 be $N_q\left(\beta_0, \Sigma_0\right)$ and $IG(\alpha_0, \beta_0)$, respectively. We will use the common notation for an inverse-gamma distribution, so that if $X \sim IG(\alpha, \beta)$ then the density is given by

$$f_X(x|\alpha, \beta) = \frac{\exp(-\beta^{-1}/x)}{\Gamma(\alpha)\beta^\alpha x^{\alpha+1}}, \qquad x \in \mathbb{R}^+,$$

where $\alpha, \beta > 0$. Then the posterior distribution $\pi(\eta|z)$ is proportional to

$$IG\left(\sigma^2|\alpha_{\sigma^2}, \beta_{\sigma^2}\right)\Phi\left(\varphi\right)|K_\theta|^{-\frac{1}{2}}\pi(\beta)\pi(\theta)\pi(\alpha), \qquad (9.6)$$

where $\alpha_{\sigma^2} = \frac{n}{2} + \alpha_0$ and

$$\beta_{\sigma^2} = \frac{2\beta_0}{(z - F\beta)'K_\theta^{-1}(z - F\beta)\beta_0 + 2}.$$

Furthermore, full conditional distributions are given by

$$\pi(\beta|z, \sigma^2, \theta, \alpha) \;\propto\; N_q\left(\mu_\beta, \Sigma_\beta\right)\Phi\left(\varphi\right), \qquad (9.7)$$

where $\mu_\beta = \Sigma_\beta\left(\frac{F'K_\theta^{-1}F}{\sigma^2}\hat\beta + \Sigma_0^{-1}\beta_0\right)$, $\Sigma_\beta = \left(\frac{F'K_\theta^{-1}F}{\sigma^2} + \Sigma_0^{-1}\right)^{-1}$, and $\hat\beta = (F'K_\theta^{-1}F)^{-1}F'K_\theta^{-1}z$,

$$\pi(\sigma^2|z, \beta, \theta, \alpha) \;\propto\; IG\left(\alpha_{\sigma^2}, \beta_{\sigma^2}\right)\Phi\left(\varphi\right). \qquad (9.8)$$

$$\pi(\theta|z, \beta, \sigma^2, \alpha) \;\propto\; |K_\theta|^{-\frac{1}{2}}exp\left(-\frac{1}{2\sigma^2}(z - F\beta)'K_\theta^{-1}(z - F\beta)\right)$$

$$\times \Phi(\varphi) \pi(\boldsymbol{\theta}). \tag{9.9}$$

$$\pi(\alpha | \boldsymbol{z}, \boldsymbol{\beta}, \sigma^2, \boldsymbol{\theta}) \quad \propto \quad \Phi(\varphi) \pi(\alpha). \tag{9.10}$$

We can modify the prior distribution of α using mixture distributions by adding additional hyperpriors taking

$$\pi(\alpha) = \pi(\alpha | \gamma, p) \pi(\gamma | p) \pi(p).$$

The conditional prior of α is $\pi(\alpha | \gamma, p) = \pi(\alpha | \gamma) = 0$ if $\gamma = 0$ and equal to $N(0, \sigma_0^2) \Phi(\varphi)$, otherwise. Further, we assume $\gamma | p \sim Ber(p)$, $p \sim Beta(\alpha_0, \alpha_0)$ so $E(p) = \frac{1}{2}$. We will use $Ber(p)$ to denote the Bernoulli distribution with parameter p and $Beta(\alpha, \beta)$ the Beta distribution with parameter α and β. Hence, the marginal prior distribution for α is

$$p(\alpha | p) = (1 - p) I(\alpha = 0) + p N(0, \sigma_0^2),$$

where $I(\alpha = 0)$ denotes the indicator function, and hence it is a mixture distribution. Therefore this model includes a Gaussian process as a special case and $(1 - p)$ gives us the probability that the data is generated from this Gaussian process. In this way the prediction will be made using a mixture of Gaussian and skew-Gaussian processes.

The full conditional distributions of α, γ, p are as follows, and the other full conditional distributions are exactly the same as in (9.7)-(9.9).

$$\pi(\alpha | \boldsymbol{z}, \gamma, p, \boldsymbol{\beta}, \sigma^2, \boldsymbol{\theta}) \quad \propto \quad \pi(\alpha | \gamma) f(\boldsymbol{z} | \boldsymbol{\eta}), \tag{9.11}$$

$$\pi(\gamma | \boldsymbol{z}, \alpha, p, \boldsymbol{\beta}, \sigma^2, \boldsymbol{\theta}) \quad \propto \quad \pi(\alpha | \gamma) \pi(\gamma | p) \propto Ber\left(\frac{a}{a+b}\right), \tag{9.12}$$

where $a = 1 - p$ and $b = (p / \sqrt{2\pi\sigma_0^2}) exp\left\{-\alpha^2 / (2\sigma_0^2)\right\}$.

$$\pi(p | \boldsymbol{z}, \gamma, \alpha, \boldsymbol{\beta}, \sigma^2, \boldsymbol{\theta}) \quad \propto \quad \pi(\gamma | p) \pi(p)$$
$$\propto \quad Beta(\alpha_0 + \gamma, \alpha_0 + 1 - \gamma). \tag{9.13}$$

Note $\boldsymbol{\eta}$ contains all parameters which are $\boldsymbol{\eta} = (\boldsymbol{\beta}', \sigma^2, \boldsymbol{\theta}', \alpha)'$. Finally the Bayesian predictive distribution is given by $f(z(\boldsymbol{x}_0) | \boldsymbol{z}) \propto \int_{\boldsymbol{\eta}} f(z(\boldsymbol{x}_0) | \boldsymbol{\eta}, \boldsymbol{z}) \pi(\boldsymbol{\eta} | \boldsymbol{z}) d\boldsymbol{\eta}$.

9.2.3 Computational Strategy

The posterior distribution is in a complicated form so we will use a Markov chain Monte Carlo (MCMC) technique (Gilks et al. 1996) to generate samples from the posterior distribution.

To generate samples from the posterior distribution $\pi(\boldsymbol{\eta} | \boldsymbol{z})$ where $\boldsymbol{\eta} = (\boldsymbol{\beta}', \sigma^2, \boldsymbol{\theta}', \alpha)'$ we will exploit the full conditional distributions, as usually done in Gibbs sampling (Gelfand and Smith 1990). Since all the five full conditional distributions except $\pi(\boldsymbol{\theta} | \boldsymbol{z}, \boldsymbol{\beta}, \sigma^2, \alpha)$ can be written as $\pi(\cdot) \propto \psi(\cdot) \Phi(\cdot)$, where $\psi(\cdot)$ is a density that can be sampled and $\Phi(\ )$ is uniformly bounded, the probability of a move in the Metropolis-Hastings algorithm requires only the computation of the $\Phi(\cdot)$ function and is given by

$$\alpha(x,y) = \min\left(\frac{\Phi(y)}{\Phi(x)}, 1\right). \tag{9.14}$$

This is an efficient solution when available (Chib and Greenberg 1995). The probability of a move for $\boldsymbol{\theta}$ is

$$\alpha(x,y) = \min\left(\frac{|K_y|^{-\frac{1}{2}}\exp\left(-\frac{1}{2\sigma^2}(\boldsymbol{z}-F\boldsymbol{\beta})'K_y^{-1}(\boldsymbol{z}-F\boldsymbol{\beta})\right)\Phi(\varphi_y)}{|K_x|^{-\frac{1}{2}}\exp\left(-\frac{1}{2\sigma^2}(\boldsymbol{z}-F\boldsymbol{\beta})'K_x^{-1}(\boldsymbol{z}-F\boldsymbol{\beta})\right)\Phi(\varphi_x)}, 1\right),$$

where φ_x (φ_y) is the same as φ having x (y) as an argument which is defined as $\varphi = \frac{\alpha}{\sigma}\alpha_z'(\boldsymbol{z}-F\boldsymbol{\beta})$. To sample from the conditional distribution, $f(z(\boldsymbol{x}_0)|\boldsymbol{z},\boldsymbol{\eta})$, we introduce a truncated normal distribution (Robert, 1995). Finally we consider a generated value of $f(z(\boldsymbol{x}_0)|\boldsymbol{z},\boldsymbol{\eta})$ as a sample from a predictive distribution by the composition method (Gelfand 1996, Tanner 1996).

The techniques for mixture prior are similar to the previously discussed methods except we need additional steps to generate α and the corresponding hyperparameters. To generate α,γ,p from the full conditional distributions, we first generate γ. If γ equals 0 then α equals 0 (likelihood $f(\boldsymbol{z}|\boldsymbol{\eta})$ is a Gaussian process), and we generate samples from $\pi(\alpha)\Phi(\varphi)$ otherwise. Sampling from γ,p is direct since we know the exact conditional distributions.

Another possible way to generate samples from a predictive distribution is a Monte Carlo integration (Gelfand 1996, Tanner 1996, De Oliveira et al. 1997).

9.3 Real Data Illustration: Spatial Potential Data Prediction

The dataset analyzed in this section is flow potential data for the North Burbank Field in California. The flow potential is a very important factor in determining flow velocity through porous media in the reservoir. For instance, in a reservoir with high flow potential, the fluid velocity (of oil/water/gas) will be fast, compared to other low flow potential reservoirs. Usually petroleum engineers want to produce oil in the reservoir as fast as possible because they want to obtain maximum profits through oil production from the reservoir, keeping the reservoir pressure. The original data have been modified slightly (for confidentiality reasons) by multiplying by, and adding, unspecified constants.

A graphical description of the locations of the flow potential data is shown in Fig. 9.1, with a histogram of the response values, ignoring position, given in Fig. 9.2. It looks like the dataset is generated from a skewed distribution, so use of the skew-Gaussian process seems appropriate here.

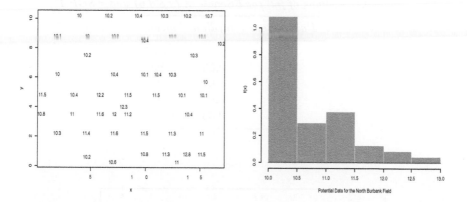

Figure 9.1 *The 48 observed locations of the flow potential data, including the response values.*

Figure 9.2 *A histogram of the observed responses.*

Another possible family relates to lognormal kriging, which was commented on earlier in Section 9.2. However, we find that the log transformed data do not follow a normal distribution, so it is not a good idea to use lognormal kriging here. In the absence of regression parameter information the mean function is assumed to be constant, so $q = 1$. In our practical usage, we adopt the general exponential isotropic correlation function (Yaglom 1987). In the following we use the second parameterization in (9.3) since it eases the assignment of non-informative priors and the MCMC methods. This is a flexible family containing the exponential ($\theta_2 = 1$) and the squared exponential ($\theta_2 = 2$) correlation functions as two of its members. It is easy to compute and is parameterized by physically interpretable quantities. For this family θ_1 controls the range of correlation, viewed as the correlation between two observations and θ_2 controls the smoothness of the random field. Another flexible family is the Matérn class of correlation functions used by Handcock and Wallis (1993) and Handcock and Stein (1994).

The first priors (SGP 1) are as follows. We assume $\beta \sim N(\bar{z}, 100)$, $\sigma^2 \sim IG(0.05, 0.05)$, $\theta_1 \sim U(0, 1), \theta_2 \sim U(0, 2]$, and $\alpha \sim N(0, 100)$ which are all independent. Hence this model also contains a Gaussian process as a special case because the values of α can be 0. Since there is no information on θ_1 and θ_2 we assume uniform priors. Applying the MCMC method of Section 9.2.3, we get full conditional distributions. We summarize medians of full conditionals in Table 9.1.

After getting full conditionals, we obtain values of each predictive density function for $Z(x_0)$ for the locations $x_0 = (2.5, 2.5), (14.5, 5.5), (14, 9), (14, 2), (6, 9.5)$ and $(7.8, 3.6)$ covering different sections of the region of in-

Table 9.1 *Medians of full conditionals at (7.8,3.6) under SGP 1*

Parameters	median	mean
β	10.32	10.40
$1/\sigma^2$	0.03	0.03
α	1.07	1.21
θ_1	0.98	0.97
θ_2	0.33	0.34

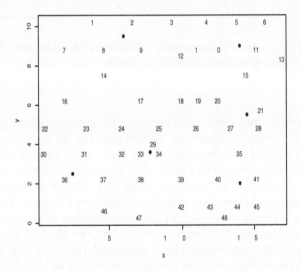

Figure 9.3 *Predicted positions with numbering 48 observed locations.*

terest. This is shown at Fig. 9.3. The predicted positions (marked with "•") with numbering 48 locations are plotted at Fig 9.3. We ordered 48 locations in an arbitrary manner resulting in no effect on the analysis.

The second priors (SGP 2) are only different at generating the values of α. That is, $P \sim Beta(1,1), \gamma|p \sim Ber(p)$ and $\alpha|p \sim (1-p)I(\alpha = 0) + p\,N(0,100)$. Notice that the mean of P is $\frac{1}{2}$, which means a priori skewness or not is equally probable. Using similar generating technique at the first priors, we get full conditional distributions. If p is close to 1, then the distribution of data is strictly skewed. If p is close to 0, then it is almost symmetric. We also get predictive distributions for the same locations as SGP 1 and the results give us indistinguishable predictive inference. Table 9.2 summarizes prediction intervals and medians for SGP 1 model. We used

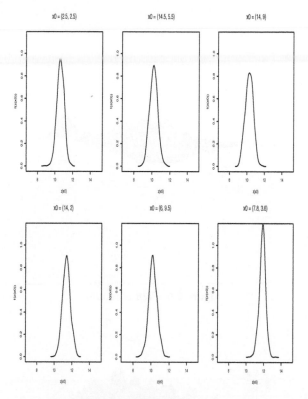

Figure 9.4 *Prediction densities for SGP 1.*

the 95% central quantile intervals since the predictive densities are pretty symmetric. Figure 9.5 shows the predicted map obtained by computing $\hat{Z}(\mathbf{s}_0) = $ median of $(Z(\mathbf{s}_0)|\mathbf{Z})$ on every $\mathbf{s}_0$ of a 40×40 grid. As a measure of predictive uncertainty at any location $\mathbf{s}_0$ we use MAD (Median Absolute Deviation) which is equal to SIQR (Semi Inter Quartile Range) and .6745 σ for a normal distribution. Figure 9.6 shows the map of this uncertainty measure.

9.4 Discussion

We used the skew-Gaussian process for the spatial prediction of flow potential data. From the final predicted map given at 9.5, we can find where the highest flow potential is located. This is an important factor in determining flow velocity through porous media in the reservoir. It performed adequately and can be used to model skewed data for which lognormal kriging cannot be used. So our method is a generalization of lognormal

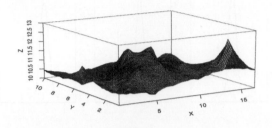

Figure 9.5 *Predicted map.*

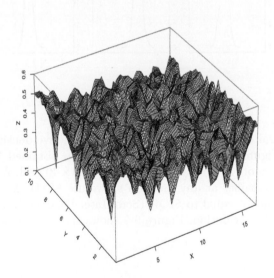

Figure 9.6 *Uncertainty map related to the predicted map.*

Table 9.2 *95% Prediction intervals and medians.*

| Location | SGP 1 | |
	P.I.	median
(2.5, 2.5)	(9.67, 11.64)	10.65
(14.5, 5.5)	(9.18, 11.22)	10.18
(14, 9)	(9.23, 11.26)	10.23
(14, 2)	(10.33, 12.30)	11.32
(6, 9.5)	(9.16, 11.25)	10.16
(7.8, 3.6)	(10.95, 12.66)	11.83

kriging in some sense. Furthermore it extends the results of usual Gaussian random fields which are a particular case of it. It can also test for presence of skewness in the data.

In Section 9.2, we chose one shape (skewness) parameter α and one scale parameter σ. Instead of those, we can consider a hierarchical model with shrinkage effect for the shape parameter vector and a diagonal matrix of scale parameters. To extend this idea the shape parameter vector and the diagonal matrix of scale parameter can be regarded as a function of distance, which is assumed at usual second order stationary condition.

Acknowledgments

We thank H.J. Newton for helping us to write the MCMC programs.

CHAPTER 10

Accounting for Absorption Lines in Images Obtained with the Chandra X-ray Observatory

D.A. van Dyk   C.M. Hans

10.1 Statistical Challenges of the Chandra X-ray Observatory

In recent years, new telescopes have dramatically improved the ability of astrophysicists to image X-ray sources. The Chandra X-ray Observatory, for example, was launched by the space shuttle *Columbia* in July 1999 and provides a high resolution tool that produces images at least thirty times sharper than any previous X-ray telescope. The X-rays themselves are produced by matter that is heated to millions of degrees, e.g., by high magnetic fields, extreme gravity, or explosive forces. Thus, the images provided by such high resolution instruments help astrophysicists to understand the hot and turbulent regions of the universe.

Unlocking the information in these images, however, requires subtle analysis. The detectors aboard Chandra collect data on each X-ray photon that arrives at the detector. Specifically, the (two-dimensional) sky coordinates, the energy, and the time of arrival of each photon are recorded. Because of instrumental constraints each of these four variables is discretized; the high resolution of Chandra means that this discretization is much finer than what was previously available. For example, one of the instruments aboard Chandra has 4096 × 4096 spatial pixels and 1024 energy bins. Because of the discrete nature of the data, it can be compiled into a four-way table of photon counts. We refer to this four dimensional data as the image; it is a moving 'colored' picture. (Because of the high energy of X-rays the 'colors' are not in the visible spectrum.) Spectral analysis models the one-way marginal table of the energy data; spatial analysis models the two-way marginal table of sky coordinates; and timing analysis models the one-way marginal table of arrival times. As we shall see, however, because of subtleties in the instrumentation and the data itself, spatial and spectral analysis cannot be fully separated and both are crucial for a full understanding of the image; in other settings, similar concerns arise for spatial and temporal analysis (see Chapter 12).

The data gathered by Chandra, although high-resolution, present a number of statistical challenges to the astronomer. For spatial data, the images are convolved with instrument characteristics which blur the image. For example, the point spread function characterizes the probability distribution of a photon's recorded pixel location relative to its actual sky coordinates. The situation is further complicated because the shape of the scatter distribution varies across the detector; it is symmetric and relatively tight in the center and becomes more asymmetric, irregular, and diffuse towards the edge. Moreover, the scatter distribution can vary with the energy of the incoming photon. Like the sky coordinates the energy of a photon is subject to "blurring"; there is a distribution of potential recorded energies given the actual energy of a particular photon. Thus, we have three dimensional blurring of the image. Given the sky coordinates and energy of a photon, there is a distribution of the recorded sky coordinates and recorded energy of the photon. Since there are, for example, 4096×4096 pixels on the detector and 1024 energy bins, the resulting blurring matrix can have over 2.9×10^{20} cells. Clearly some simplification is required. For spectral analysis using a small region of the detector, the blurring of energies is more-or-less constant, which results in a reasonably sized (1024×1024) matrix. Thus, utilizing sparse matrix techniques results in efficient computation for marginal spectral analysis. Spatial analysis often involves only a subset of the pixels, reducing the dimension of the problem. Also the blurring matrix can be taken to be constant across a large number of pixels and energy bins. Thus, we might divide the energy bins into 4 groups and the pixels into 16 groups and assume the shape of the energy cross sky coordinate scatter is constant in each of the resulting 64 cells. Such techniques aim at computational efficiency while hoping the compromise in precision is minor. A careful analysis of this trade off has yet to be tackled.

Another complication for image analysis involves the absorption of photons and the so-called effective area of the telescope. Depending on the energy of a photon, it has a certain probability of being absorbed, for example by the inter-stellar media between the source and the detector. Effective area is an instrumental effect, but has a similar effect on the data—the probability that a photon is recorded by the detector depends on its energy. Because the spectrum can vary across the source, the rate of this stochastic censoring also varies and can distort the image. Again, this emphasizes that a careful analysis of the image must involve spectral analysis. In this paper, we take up the task of modeling photon absorption—with the understanding that it has direct implications for spatial analysis. In particular, we introduce new models that aim to account for absorption lines in spectra. An absorption line is a narrow range of energies where a relatively high proportion of photons are absorbed. Because these lines are caused by an abundance of a particular element near the surface of the source, it is possible for the intensity of an absorption line to vary across the source, thus distorting the image.

The data are also degraded by the presence of background counts, X-ray photons which arrive at the detector but do not correspond to the source of interest. In spectral analysis, a second observation that consists only of background counts is compared with the primary observation. The background observation is obtained by looking into an area of space near the source but which contains no apparent X-ray sources. After adjusting for exposure time, in some analyses the background observation is subtracted from the source observation, and the result is analyzed as if it were a source observation free of background. This procedure is clearly questionable, especially when the number of counts per bin is small. It often leads to the rather embarrassing problem of negative counts and can have unpredictable results on statistical inference. A better strategy is to model the counts in the two observations as independent Poisson random variables, one with only a background intensity and the other with intensity equal to the sum of the background and source intensities (Loredo 1993, van Dyk et al. 2001).

A final degradation of the data is known as pile up and poses a particularly challenging statistical problem. Pile-up occurs in X-ray detectors (generally charged coupled devices, i.e., CCDs) when two or more photons arrive at the same location in the detector (i.e., in an event detection island, which consists of several pixels) during the same time bin. Such coincident events are counted as a single higher energy event or lost altogether if the total energy goes above the on-board discriminators. Thus, for bright sources pile-up can seriously distort the count rate, the spectrum, and the image. Moreover accounting for pile up is inherently a task of joint spectral-spatial modeling. A diffuse extended source may have no appreciable pile up because the count rate is low on any one area of the detector. A point source with the same marginal intensity, however, may be subject to severe pile up. Model based methods for handling pile up are discussed in Kang et al. (2002); see also Davis (2001).

We propose using model-based Bayesian methods to handle these complex imaging problems; other Bayesian approaches to image analysis appear, for example, in Chapters 6, 7, 8, 11 and 14 of this volume. Models can be designed to handle not only the complexity of the data collection process (e.g., blurring, effective area of the instrument, background contamination, and pile up) but also the complex spatial and spectral structures of the sources themselves. For example, as discussed in Section 10.2, we are interested in clustering photons into spatial and spectral features of the source. Because of their complexity, the models are in turn complex and require sophisticated computational methods for fitting. A Bayesian perspective is ideally suited to such highly structured models in terms of both inference and computation. For example, the high dimensional parameter space and numerous nuisance parameters highlight the attraction of Bayesian marginal posterior distributions of parameters or groups of parameters. Astrophysicists are often interested in testing for the presence of a particular model feature. Many such tests, such as testing for the presence of an additional point source in an image, correspond to testing whether a parameter is on the boundary of its space. It is well known that the likelihood ratio and related tests fail in this setting. However, appropriate Bayesian model fitting and model checking procedures are readily available (e.g. Protassov et al. 2001).

From a computational point of view, such tools as the EM algorithm (Dempster et al. 1977a), the Data Augmentation Algorithm (Tanner and Wong 1987), the Gibbs sampler (e.g. Gelfand and Smith 1990, Smith and Roberts 1993), and other Markov chain Monte Carlo methods are ideally suited to highly structured models of this sort; see van Dyk (2002). The modular structure of these algorithms matches the hierarchical structure of our models. For example, the Gibbs sampler samples one set of model parameters from their conditional posterior distribution given all other model parameters. This allows us to fit one component of the overall model at a time, conditional on the others. Thus, a complex model fitting task is divided into a sequence of much easier tasks. This modular structure also allows us to take advantage of well known algorithms that exist for fitting certain components of our model. For example, using the EM algorithm to handle a blurring matrix and background contamination in Poisson image analysis is a well known (and often rediscovered) technique (Richardson 1972, Lucy 1974, Shepp and Vardi 1982, Lange and Carson 1984, Fessler and Hero 1994, Meng and van Dyk 1997). Even though this standby image reconstruction algorithm is unable to handle the richness of our highly structured model, we utilize it and its stochastic generalization as a step in our mode finding and posterior sampling algorithms.

Detailing how we (or how we expect to) handle all of the modeling, computational, and inferential aspects of image analysis of Chandra data is well beyond the scope of this chapter. Instead, we outline some of our models to give the reader a flavor of our Bayesian analysis and highly structured models. In some cases, the details can be found in one of several references; in other cases, methods are still being developed. To give the reader a flavor of the statistical details that are involved, however, we go into some depth in our description of absorption lines.

The remainder of this chapter is organized into four sections. In Section 10.2 we outline our marginal spatial and spectral models, paying particular attention to the model based clustering. We discuss absorption lines in Section 10.3, describing the science behind them, the models and computational methods we propose, and a simulation study which illustrates some of the statistical difficulties involved. In Section 10.4, we incorporate absorption lines into the spectral model discussed in Section 10.2 illustrating our methods with a data set. Concluding remarks appear in Section 10.5.

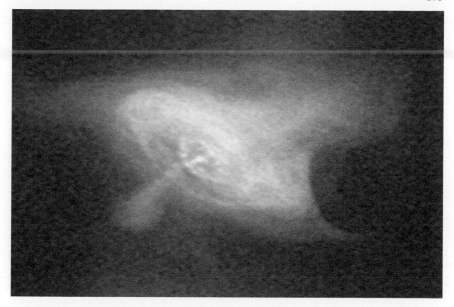

Figure 10.1 *This X-ray image of the Crab Nebula is one of the first images sent back by* Chandra. *The Crab Nebula is one of the youngest and most energetic of about 1000 known pulsars; in fact this supernova remnant produces energy at the rate of 100,000 suns. The image illustrates the extended irregular spatial structure that is typical of* Chandra *images (The image was adaptively smoothed; image credit: the National Aeronautics & Space Administration (NASA), the Chandra X-ray Center (CXC), and the Smithsonian Astrophysical Observatory (SAO)).*

10.2 Modeling the Image

10.2.1 Model-Based Spatial Analysis

We begin by modeling the true source counts in each pixel, $X = \{X_{i+}, i \in \mathcal{I}\}$, as independent Poisson random variables,

$$X_{i+} \sim Poisson(\Lambda_{i+}) \text{ for } i \in \mathcal{I}, \tag{10.1}$$

where $\mathcal{I}$ is the set of pixels and the '+' in the subscript indicates that we are summing over the time and energy bins; in the remainder of Section 10.2.1 we suppress the '+'. Because the data are degraded by factors such as image blurring and background contamination as discussed in Section 10.1, X is not observed. Thus, we discuss both the constraints on Λ that we impose to represent structure in the image and how we model data distortion. We begin with Λ.

To motivate our parameterization of Λ we examine several Chandra images. Figure 10.1 illustrates an X-ray image of the Crab Nebula, the remnant of a supernova explosion that was observed on Earth in 1054 A.D; im-

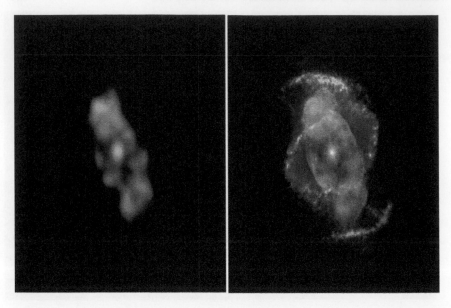

Figure 10.2 *X-ray image of the Cat's Eye Nebula (left) observed by* Chandra, *and a composite X-ray/optical observed by* Chandra *and the Hubble Space Telescope (HST) (right). The image on the left shows a bright central star, which corresponds to a high density cluster of photons. The composite image on the right illustrates the relative locations of hotter and cooler regions of the planetary nebula. (The X-ray image is adaptively smoothed. The color composite of optical and X-ray images was made by Zoltan G. Levay (Space Telescope Science Institute). The optical images were taken by J.P. Harrington and K.J. Borkowski (University of Maryland) with HST. Image credits: NASA, University of Illinois at Urbana-Champaign, Chu et al. (2001), and HST).*

age brightness corresponds to X-ray intensity. At the center of the nebula is a rapidly spinning pulsar that emits a pulse of photons thirty times per second. The Crab Nebula is a very bright X-ray source and illustrates the extended irregular structure that is typical of X-ray images. The structure in the extended source can sometimes be predicted from optical or radio images but often contains unique features. For example, the jet that extends towards the lower left from the center of the nebula was first observed by Chandra. Although model based methods are not required to identify some of the important structures in the Crab Nebula, such methods have broad application for analyzing weaker X-ray sources, understanding how the energy spectrum varies across a source, and identifying weak features in the source.

A second image appears in Figure 10.2 and illustrates X-ray (left panel) and optical (right panel) images of the Cat's Eye Nebula. A bright central

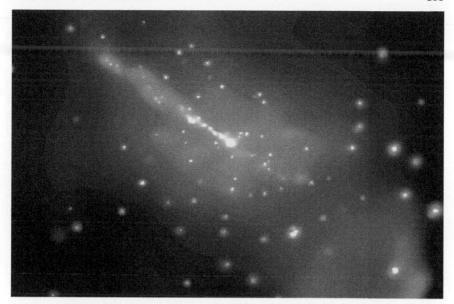

Figure 10.3 *Centaurus A, a nearby galaxy, as observed by* Chandra. *The several features of the galaxy – a super-massive black-hole, a jet emanating from the core and the point-like sources scattered about the image – have been clearly resolved for the first time due to the imaging resolution of* Chandra *and illustrate the variety of photon clusters that might be present in a* Chandra *image. (The image is adaptively smoothed and exposure corrected; image credit: NASA, SAO, and Kraft et al. (2001)).*

star is apparent in the center of the multi-million-degree nebula. Again the nebula exhibits extended irregular structure that is only partially predictable from the optical image. A final image appears in Figure 10.3, this one of a nearby elliptical galaxy, Centaurus A. The image shows a bright central source, which is believed to include a super-massive black hole, a jet emanating from the center, and numerous point-like X-ray sources all surrounded by a diffuse hot gas.

Figures 10.1-10.3 illustrate the wide variety of images that we hope to model. A useful model must both allow for the extended diffuse nebula with its irregular and unpredictable structure and include one or more highly concentrated X-ray emitter (i.e., point sources). An important objective is to cluster photons into these various sources. To accomplish this, we model X_i as a mixture of independent Poisson random variables. In particular,

$$\Lambda_i = \lambda_i^{\mathrm{ES}} + \sum_{k=1}^{K} \lambda_k^{\mathrm{PS}} p_{ik} \text{ for } i \in \mathcal{I}, \tag{10.2}$$

where λ_i^{ES} represents the expected count due to smooth irregular extended source in pixel i, K is the number of point sources, λ_k^{PS} is the total expected count due to point source k, and p_{ik} is the proportion of point source k that falls in pixel i. The added point sources might be modeled as Gaussian distributions, in which case p_{ik} is a bivariate Gaussian integral. We can easily handle larger elliptical Gaussian point sources, or more irregular point source models. Information as to the location and spread of the point source is often forthcoming (e.g., from optical or radio observations) and thus informative prior distributions are often available for these parameters.

A Markov Random Field (MRF) model can be used to describe the irregular extended sources represented by λ_i^{ES} in (10.2) (and perhaps irregular added point sources). Our MRF models each pixel's log intensity as a Gaussian deviate centered at the mean of the log intensities of its neighboring pixels. In particular,

$$\log(\lambda_i^{\mathrm{ES}}) \sim \mathrm{Normal}\left(\frac{1}{n_i} \sum_{j \in \partial(i)} \log(\lambda_j^{\mathrm{ES}}), \frac{v}{n_i}\right) \text{ for } i \in \mathcal{I}, \qquad (10.3)$$

where $\partial(i)$ is the set of pixels neighboring pixel i, n_i is the number of pixels in $\partial(i)$, and v is the user-specified between pixel variance. This specification allows for flexibility in defining each pixel's neighborhood, as well as specifying the variance of the Gaussian density. In principle, the variance parameter can be fit or can be allowed to vary across the detector field, so as to give the capacity for sharp edges. The variances themselves can then be modeled, perhaps via a common prior distribution with fitted hyperparameters, letting the data in effect determine the values of the smoothing parameter. An alternative to the MRF is the multiscale method proposed by Kolaczyk (1999), which aims to provide wavelet like models for Poisson data and has performed well in a variety of applications (Nowak and Kolaczyk 2000); the methods described in Chapters 6 and 7 may also be helpful in identifying regions of relative homogeneity in an extended source.

We turn now to models for the degraded observed data. As discussed in Section 10.1, the observed counts are blurred because of instrumental effects and contaminated by background counts. Thus, we modify (10.2) to model the observed count in pixel i, Y_i, as independent Poisson random variables,

$$Y_i \sim Poisson\left(\sum_{j \in \mathcal{I}} M_{ij}\Lambda_j + \theta_i^{\mathrm{B}}\right) \text{ for } i \in \mathcal{I}, \qquad (10.4)$$

where M_{ij} is the probability that a photon with actual sky coordinates corresponding to pixel j is recorded in pixel i, Λ_j is given in (10.2), and θ_i^{B} the expected counts in pixel i attributed to background contamination. The blurring matrix (M_{ij}) and generally the background vector (θ_i^{B}) are assumed known from calibration. As discussed in Section 10.1, the model given in (10.4) is well known and has been the subject of much study. What is new here are the constraints we put on the source model and the spectral models to which we now turn.

10.2.2 Model-Based Spectral Analysis

In this section we briefly outline a class of spectral models; more details of the models and the algorithms used to fit them can be found in van Dyk et al. (2001), Protassov et al. (2001), and Sourlas et al. (2002). The spectral model aims to describe the distribution of the photon energies emanating from a particular source. Generally speaking, the distribution consists of several clusters of photons including a smooth *continuum* term and a number of added *emission lines*, which are narrow ranges of energy with more counts than would be expected from the continuum. The continuum is formed by radiation of heat from the hot center of stars to the cold space that surrounds them, a process known as thermal Bremsstrahlung. The emission lines are due to particular ions in the source and the abundance of the extra emission indicates the abundance of the ion in the source. Known spectral lines can tell us about the speed at which the source is moving by examining the Doppler shift of the line location.

Statistically, the models are designed to summarize the relative frequency of the energy of photons arriving at the detector and to separate the photons into clusters corresponding to the continuum and emission lines. Independent Poisson distributions are more appropriate to model the counts than the commonly used normal approximation (e.g., χ^2 fitting), especially for a high resolution detector. We parameterize the intensity in bin $j \in \mathcal{J} = \{1, \ldots, J\}$, as a mixture of the continuum term and K emission lines,

$$\Lambda_j = \delta_j \lambda^{\mathrm{C}}(\theta^{\mathrm{C}}, E_j) + \sum_{k=1}^{K} \lambda_k^{\mathrm{E}} p_{jk}, \quad j \in \mathcal{J}, \tag{10.5}$$

where δ_j is the width of bin j, $\lambda^{\mathrm{C}}(\theta^{\mathrm{C}}, E_j)$ is the continuum term and is a function of the continuum parameter, θ^{C}, E_j is the mean energy in bin j, λ_k^{E} is the expected counts from emission line k, and p_{jk} is the proportion of emission line k that falls in bin j. (Here and in the rest of the chapter we refer to the spectral margin, although this is suppressed in the notation.) The smooth continuum term generally is parameterized according to physical models with several free parameters. Many of these models are amenable to standard statistical techniques, e.g., log linear models. Occasionally, a less parametric fit is desired, in which case a one dimensional Markov random field can be applied. The emission lines are generally parameterized as Gaussian or t distributions.

As discussed in Section 10.1, the counts are degraded by background contamination, instrument (i.e., the detector) response, and photon absorption. Instrument response is a characteristic of the detector that results in a blurring of the photons, i.e., a photon that arrives in bin j has probability M_{lj} of being detected in observed bin $l \in \mathcal{L} = \{1, \ldots, L\}$. The $L \times J$ matrix $\{M_{lj}\}$, which may not be square, is determined by calibration of the detector and presumed known. Because of these degradations, we model the observed counts as independent Poisson variables with parameters

$$\Xi_l = \sum_{j=1}^{J} M_{lj} \Lambda_j d_j \alpha(\theta^{\mathrm{A}}, E_j) + \theta_l^{\mathrm{B}}, \quad l \in \mathcal{L}, \qquad (10.6)$$

where $\alpha(\theta^{\mathrm{A}}, E_j)$ is the probability that a photon of energy E_j is *not* absorbed, θ_l^{B} is the Poisson intensity of the background which is often known from calibration in space, and d_j is the known effective area for photons with energy E_j, with d_j normalized so that $\max_j d_j = 1$. In the absorption model $\alpha(\theta^{\mathrm{A}}, E_j)$ is typically taken to be a generalized linear model with θ^{A} denoting the model parameter. The absorbed photons form one or more clusters that are completely unobserved. An important special case involves so-called absorption lines, which are the topic of the remainder of the chapter.

10.3 Absorption Lines

10.3.1 Scientific Background

Absorption lines are downward spikes in the spectral continuum, which represent wavelengths where photons from the continuum have been absorbed by atoms of elements in the source. Because the specific energies at which photons are absorbed are unique to each element, examining absorption lines of a source can help to determine its composition. In order to motivate the physical models we employ and the statistical models we formulate, we begin with some scientific background. Photons are emitted from the hot center of a source (e.g., a star) in a continuous spectrum, and due to their high energy radiate toward the relatively colder region near the "surface" of the source (e.g., the corona of a star). In these cooler regions the continuum photons are in a higher energy state than their surroundings and thus are readily absorbed by surrounding atoms to keep the energy of the system in balance. When this occurs, the absorbing atom necessarily enters a higher, less stable, energy state. Any given atom, however, prefers a lower, more stable energy configuration, so with high likelihood the atom will shed the excess energy and return to its original state.

If the energy of the absorbed photon is eventually released by the absorbing atom, one may wonder why we observe a dip in the continuum. The answer is in part due to two processes called collisional deexcitation and radiative deexcitation. In collisional deexcitation the excited atom collides with another atom, and the "extra" energy due to the absorbed photon is converted into kinetic energy. The previously excited atom is thus back to its original state, and we still observe a downward spike in the continuum. In radiative deexcitation the atom simply emits a photon of energy equal to the absorbed photon. However, the chance that we observe this photon with the detector is very small: there are many possible directions which the emitted photon can take, and the probability is minute that its path will be along our line of observation. Therefore, even though the excited atoms eventually return to lower energy states, we still observe an absorption line.

Absorption lines can be parameterized in terms of their location, width, and intensity. The location, μ, denotes the center of the absorption line and is of interest because the absorption wavelength (or equivalently, energy) indicates the absorbing element. Absorption lines generally have some positive width, σ^2, because they are "broadened" by several effects. One of these effects, Doppler broadening, occurs because the velocity at which a photon is moving when it is absorbed is a random variable, causing us to observe a Doppler "shifted" absorption energy. This causes some photons to appear to be absorbed at a slightly higher energy and others at a slightly lower energy, hence the broadening of absorption lines. The third parameter is the intensity parameter, λ^{A}. Astronomers often refer to the absorption mechanism that produces a line with small λ^{A} as being "optically thin," which means that the line does not absorb all of the continuum photons at its peak; an example appears in plot (a) of Figure 10.4. The intensity and width parameters together give an indication of the structure of the line, from which astronomers can learn about the relative concentration of the absorbing element in the source.

10.3.2 Statistical Models

In terms of the model specification in Section 10.2.2, formulating an absorption model requires us to specify the probability that a photon is *not* absorbed as a function of energy, i.e., $\alpha(\theta^{\mathrm{A}}, E_i)$ in (10.6). Generally there may be several types of absorption that act independently, e.g., absorption by the inter-stellar media over a broad energy range along with several relatively narrow absorption features. Thus, we can specify $\alpha(\theta^{\mathrm{A}}, E_i)$ as a product,

$$\alpha(\theta^{\mathrm{A}}, E_i) = \prod_{j=1}^{J^{\mathrm{A}}} \alpha_j(\theta_j^{\mathrm{A}}, E_i), \tag{10.7}$$

where J^{A} is the number of independent absorption components.

For simplicity, we focus on the case when $J^{\mathrm{A}} = 1$ with the understanding that we can repeatedly apply the methods we describe to handle multiple components. This is a particularly useful exercise because of the modular structure of our model and fitting algorithms. For example, the Gibbs sampler fits the model one component at a time. Thus, a method for fitting the

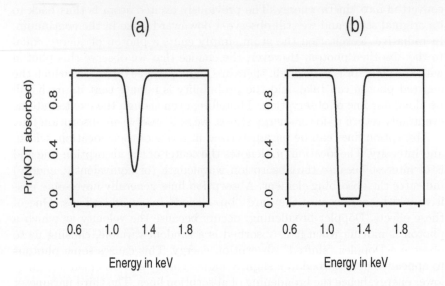

Figure 10.4 *Two examples of the physical flexibility of* $\alpha(E_j, \theta^A)$. *The vertical axis is the probability that a photon is* not *absorbed by the line, which is the probability that a photon is observed.*

spectral models described in Section 10.2.2 without absorption lines (see van Dyk et al. 2001) can be combined with the methods described here to fit the entire model, as we discuss below and illustrate in Section 10.4.

There are several models for absorption lines used by astrophysicists; some lines are modeled with flat edges on one side, and others are taken to have a symmetric or even Gaussian shape. Here, we limit our attention to a specific but important formulation of an absorption line; van Dyk et al. (2001) discuss models for absorption over a broad range of energy. Specifically, we consider the exponentiated Gaussian form described by Freeman et al. (1999),

$$\alpha(\theta^A, E_i) = \exp\left\{-\lambda^A \exp\left[\frac{-(E_i - \mu)^2}{2\sigma^2}\right]\right\}, \qquad (10.8)$$

both because it is accepted as a useful description of physical phenomena and because, as discussed is Section 10.3.3, it is computationally tractable; see also Hans and van Dyk (2002). In this model, the absorption line parameter is $\theta^A = (\mu, \sigma^2, \lambda^A)$, where μ is the location parameter, σ^2 is the width parameter and λ^A is the intensity parameter. This parameterization is attractive for both physical and, as discussed below, statistical reasons. From a physical perspective, (10.8) provides a flexible way to describe the absorption line. The inner exponential quantity gives the line a somewhat Gaussian shape, while the outer exponential and the intensity parameter control the strength of the line. Figure 10.1 illustrates this flexibility. Notice that in plot (a), the absorption line maintains a Gaussian shape but never reaches $\alpha(\theta^A, E_i) = 0$, where all the photons from the continuum would be absorbed. For this plot, $\mu = 1.25$, $\sigma^2 = 0.002$ and $\lambda^A = 1.5$. Plot (b) shows the outer limits of the line maintaining a Gaussian shape while all of the photons from the continuum are absorbed over the central bins; here, $\mu = 1.25$, $\sigma^2 = 0.002$ and $\lambda^A = 85$.

We specify the likelihood of the observed counts as

$$Y_l \overset{\text{indep.}}{\sim} Poisson(\Xi_l) \text{ for } l \in \mathcal{L}, \tag{10.9}$$

with Ξ_l given in (10.6). (Currently, it is common practice to account for absorption lines by modeling Y_l as independent Gaussian random variables with mean Ξ_l, for example, via χ^2 fitting. Such models are inappropriate for high-resolution, low-count detectors.) To complete the specification of the model, prior distributions must be assigned to the absorption line parameters. When prior information is available either from previous observations or other scientific knowledge, we use parameterized independent prior distributions on $(\sigma^2, \mu, \lambda^A)$; in particular we use scaled inverse χ^2, Gaussian, and gamma distributions respectively. Improper prior distributions should only be used with great care; there is a possibility of an improper posterior distribution when we consider the more general model described in Section 10.2.2.

Data Augmentation. We can augment model (10.9) to a larger, only partially observed sample space, which simplifies computation. The basic data augmentation is the idealized image that is undistorted by blurring, background contamination, or absorption,

$$X_i \overset{\text{indep.}}{\sim} Poisson(\Lambda_i) \text{ for } i \in \mathcal{I}, \tag{10.10}$$

with Λ_i given in (10.5). To account for absorption, we introduce an intermediate data augmentation, the idealized image *after absorption*,

$$Z_i | X_i, \theta^A \overset{\text{indep.}}{\sim} Binomial\left[X_i, \alpha(\theta^A, E_i)\right] \text{ for } i \in \mathcal{I}, \tag{10.11}$$

where $\alpha(\theta^A, E_i)$ is the probability that a photon in not absorbed and is given in (10.8). Combining (10.10) with (10.11) and marginalizing over X_i yields

$$Z_i | \theta^A \overset{\text{indep.}}{\sim} Poisson[\Lambda_i \alpha(\theta^A, E_i)] \text{ for } i \in \mathcal{I}. \tag{10.12}$$

Ordinarily we treat both $X = \{X_i, i \in \mathcal{I}\}$ and $Z = \{Z_i, i \in \mathcal{I}\}$ as missing data, along with a number of other quantities; see van Dyk et al. (2001). For the remainder of Section 10.3, however, we focus attention on absorption lines and treat Z as observed data and X as the unobserved idealized image. In particular, we act as if there were no blurring of photon energies or background contamination and assume $\Lambda = \{\Lambda_i, i \in \mathcal{I}\}$ is specified with no unknown parameters. All of these simplifications are to focus attention on absorption lines and will be relaxed in Section 10.4.

A Generalized Linear Model. The model specified in (10.11) can be formulated as a generalized linear model (McCullagh and Nelder 1989) using the link function, $\eta_i = -\log[-\log \alpha(\theta^A, E_i)]$, where α is given in (10.8). In this case,

$$\eta_i = -\log\left(\Lambda^A\right) + \frac{(E_i - \mu)^2}{2\sigma^2} = \left[-\log\left(\lambda^A\right) + \frac{\mu^2}{2\sigma^2}\right] + \left[-\frac{\mu}{\sigma^2}\right]E_i + \left[\frac{1}{2\sigma^2}\right]E_i^2$$
$$(10.13)$$

is linear in E_i and E_i^2. We can identify the coefficients of the generalized linear model with $\beta = (\beta_0, \beta_1, \beta_2)^\top$, via the invertible transformation

$$\beta_2 = \frac{1}{2\sigma^2}, \quad \beta_1 = -\frac{\mu}{\sigma^2}, \quad \beta_0 = -\log\left(\lambda^A\right) + \frac{\mu^2}{2\sigma^2}. \qquad (10.14)$$

10.3.3 Model Fitting

Our goal is to base inference on summaries of the posterior distribution,

$$p(\theta^A|Z) = \int p(\theta^A, X|Z)dX \propto \int p(Z|\theta^A, X)p(X)p(\theta^A)dX, \qquad (10.15)$$

where the factors under the final integral are given in (10.11), (10.10), and the prior distribution of θ^A respectively. Because of the complexity of (10.15) we resort to iterative methods to summarize the posterior distribution. Here we discuss both an EM algorithm that can be used to compute the posterior mode and MCMC methods that can be used to obtain a sample from the posterior distribution. Both of these methods are based on the data-augmentation scheme discussed in Section 10.3.2. In particular, both computational tools take advantage of the fact that the two conditional distributions, $p(X|Z, \theta^A)$ and $p(\theta^A|Z, X)$, are well-known statistical models. Simple probability calculus shows that the first is

$$X_i|Z_i, \theta^A \stackrel{\text{indep.}}{\sim} Z_i + Poisson\left[\lambda_i(1 - \alpha(\theta^A, E_i))\right]. \qquad (10.16)$$

The second is the posterior distribution under the generalized linear model described in Section 10.3.2.

EM Algorithm. The EM algorithm is a well-known iterative method for computing marginal posterior modes, such as the mode of $p(\theta^A|Z)$ as expressed in (10.15). Starting with some starting value $\theta^A_{(0)}$, EM proceeds by computing

$$\theta^A_{(t+1)} = \text{argmax}_{\theta^A} E\left[\log p(\theta^A|Z, X)\Big| Z, \theta^A_{(t)}\right] \text{ for } t = 1, 2, \ldots \qquad (10.17)$$

This procedure is guaranteed to increase the log posterior at each iteration and takes a particularly simple form in this case. The expectation in (10.17) can be written as

$$
E\left[\sum_{i\in\mathcal{I}} X_i \log[\alpha(\theta^A, E_i)] + (Z_i - X_i)\log[1 - \alpha(\theta^A, E_i)] + \log p(\theta^A)\Big| Z, \theta^A_{(t)}\right].
$$
(10.18)

Since (10.18) is linear in X, we can simply replace the missing data by its expectation under model (10.16) and update the parameter by fitting the generalized linear model, e.g., via Newton-Raphson; see Hans (2001) for details.

MCMC Methods. The posterior distribution in (10.15) can be summarized via Monte Carlo by obtaining a sample from $p(\theta^A, X|Z)$ and discarding the draws of X. We obtain a sample from the joint posterior distribution using the Gibbs sampler, an iterative algorithm that constructs a Markov chain which under mild regularity conditions converges to the joint posterior distribution (for convergence results see Roberts 1996).

We implement the following Gibbs sampler: given a starting value $\theta^A_{(0)}$, we iterate,

STEP1: Draw $X_{(t+1)}$ from $p(X|Z, \theta^A_{(t)})$,

STEP2: Draw $\theta^A_{(t+1)}$ from $p(\theta^A|Z, X_{(t+1)})$.

For sufficiently large T_0 we can consider $\{\theta^A_{(t)}, X_{(t)}, t = T_0, \ldots, T\}$ to be a sample from (10.15) and summarize the posterior via Monte Carlo integration.

STEP 1 in the Gibbs sampler can be easily accomplished according to (10.16). Although the probability distribution in STEP 2 is not of a standard form, we can use the Metropolis-Hastings algorithm within each iteration of the Gibbs sampler to construct a Markov chain with stationary distribution as given in STEP 2. We construct the Metropolis-Hastings jumping distribution using a wide-tailed approximation of the target density given in STEP 2. The wide tails enable the sampler to jump across the parameter space, and if the approximation is good many proposals will be accepted. Our choice for the jumping density is a multivariate location-scale t-distribution with 4 degrees of freedom. We use the posterior mode of $p(\theta^A|Z, X)$ (e.g, as computed in the M-step of EM) and the corresponding second derivative matrix to construct the center and scale of the jumping distribution respectively. Thus, the jumping distribution does not change within a single iteration of the Gibbs sampler. Because the Metropolis-Hastings algorithm is computationally quick once the jumping distribution has been computed, we iterate five times within each iteration of the Gibbs sampler, using the final draw as the draw for STEP 2 of the Gibbs sampler. This strategy has negligible costs but potentially can improve the overall convergence properties of the Markov chain.

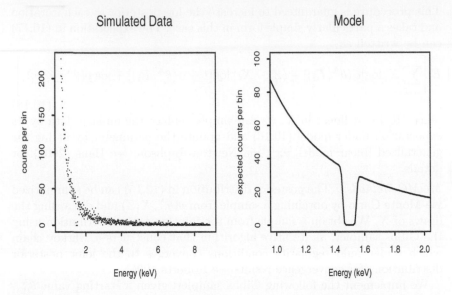

Figure 10.5 *Data simulated with* $\lambda^A = 50$. *The plot on the right shows the expected number of counts per bin in the area of the absorption line.*

10.3.4 A Simulation-Based Example

In this section we investigate how the characteristics of the fitted absorption line are affected by its actual parameters. We use a series of simulated data sets, generated according to (10.5) over the energy range [0.63 keV, 8.85 keV] using a bin width of 0.01 keV, giving a total of 822 bins. The continuum model was taken to be a power law, $\lambda^C(\theta^C, E) = \theta_1^C E^{-\theta_2^C}$, with the physically reasonable parameters $\theta_1^C = 80$ and $\theta_2^C = 2.23$. We simulated five datasets with the same line location, $\mu = 1.5$ keV, and width, $\sigma^2 = 0.0003$, but differing intensities, $\lambda^A = 1, 5, 10, 25$, and 50. The data set generated with $\lambda^A = 50$ is illustrated in Figure 10.5; the absorption line covers about ten bins. We used a flat prior distribution on θ^A in all analyses.

For each analysis we first ran the EM algorithm from five starting values dispersed about the parameter space to search for posterior modes. We then used these modes to select starting values for three MCMC samples. We begin with our analysis of the data set illustrated in Figure 10.5, generated with $\lambda^A = 50$. We fit the model to the data in two ways: first we allowed all three parameters in θ^A to be fit and second we fit only μ and λ^A, fixing σ^2 at 0.0003. The convergent values of EM for all five starting values for both fitting schemes appear in Table 10.1, which illustrates the

| Run | Fit μ, σ^2, and λ^A | | | | Fit only μ and λ^A | | |
	μ	$\sigma^2 \times 10^4$	λ^A	$\ell(\theta)$	μ	λ^A	$\ell(\theta)$
(i)	1.50	3.77	26.09	29936.3	1.30	0.23	29651.5
(ii)	1.50	3.77	26.09	29936.3	1.97	0.30	29650.6
(iii)	1.50	3.77	26.09	29936.3	1.50	52.09	29950.2
(iv)	1.50	3.77	26.09	29936.3	2.50	0.63	29654.1
(v)	1.50	3.77	26.09	29936.3	1.00	0.15	29650.5

Table 10.1 *Convergence of the EM algorithm for various starting values with $\lambda^A = 50$ for the underlying model $(\mu_{(0)}, \sigma^2_{(0)}, \lambda^A_{(0)})$ chosen as: (i) $(1.25, 5\times10^{-4}, 45)$; (ii) $(2.0, 2.0\times10^{-3}, 80)$; (iii) $(1.46, 2.5\times10^{-4}, 25)$; (iv) $(2.5, 8.0\times10^{-2}, 100)$ and (v) $(1.0, 5.0\times10^{-5}, 10)$. The left side of the table shows convergence for the model which fits all three parameters, and the right side shows convergence when σ^2 is fixed at 3×10^{-4}. The reported log-likelihood, $\ell(\theta)$, does not include the normalizing constant.*

multi-modal character of the posterior distribution. The natural Poisson variability of the photon counts can lead to random dips in the continuum which are not due to an absorption line, but create modes in the posterior distribution. Given our knowledge of the true model, the four small values of λ^A, and the value of the loglikelihood at each of these modes, it is evident that there is one major mode due to the absorption line and four minor modes that result from random fluctuations in the continuum. In practice these minor modes cause two difficulties. Computationally, MCMC samplers can get caught in a minor, relatively uninteresting mode. Thus, we generally recommend using the EM algorithm to first identify interesting modes and then construct MCMC starting values aiming to sample from these modes; see also Gelman et al. (1995). Secondly, it can be difficult to distinguish actual absorption lines from chance Poisson fluctuations in the continuum. The standard formulation of a formal hypothesis test involves a null value (i.e., no absorption line) that is on the boundary of the parameter space. Thus, the standard null distribution of the likelihood ratio test is inappropriate. In this case, we recommend using model checking techniques such as posterior predictive p-values to help distinguish between random fluctuations and weak lines; see e.g., Protassov et al. (2001).

A sample from the posterior distribution was generated by running each of three MCMC chains for 2500 iterations. We discarded the initial 500 draws of each chain and computed $\sqrt{\hat{R}}$ (Gelman and Rubin 1992) on the remaining draws to determine convergence to stationarity. Values of $\sqrt{\hat{R}}$ close to one signify that the several chains represent draws from the same distribution; values for three estimands (μ, σ^2 and $\log(\lambda^A)$) are reported in

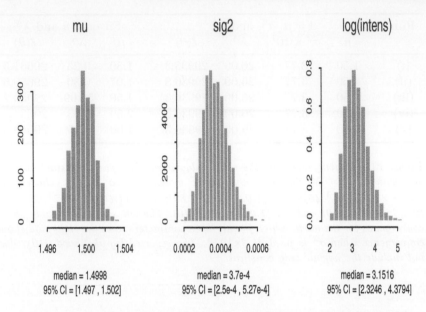

Figure 10.6 *6000 draws from the posterior distribution of θ^A for the data gener-ated with $\lambda^A = 50$; the median and 95% credible intervals are reported. On the original scale, the posterior median for the intensity parameter is 23.374 with 95% CI = [10.223 , 79.794].*

Table 10.2. The $\sqrt{\hat{R}}$ statistic and visual inspection of the chains indicate good mixing. Figure 10.6 illustrates the marginal posterior distributions of μ, σ^2, and $\log(\lambda^A)$ and reports their median and 95% credible intervals; the true value of each is contained in its interval.

Finally, we repeat the above analyses for the datasets generated with $\lambda^A = 25, 10, 5$, and 1; see Table 10.2. Figure 10.7 shows that 95% credi-ble intervals cover the true parameter value in all cases. We also notice a negative association between the estimates of σ^2 and λ^A; when σ^2 is under-estimated, λ^A is overestimated and vice versa. In either case, the expected absorption count is maintained.

10.4 Spectral Models with Absorption Lines

10.4.1 Combining Models and Algorithms

In this section we relax the model simplifications of Section 10.3.2, fitting absorption lines and the continuum jointly in the presence of background contamination, absorption due to the inter-stellar media, and blurring of photon energies. The idealized spectrum after absorption, Z, is treated as one level in a hierarchical data augmentation scheme; the observed data,

model	summary	μ	σ^2	$\log(\lambda^A)$	λ^A
$\lambda^A = 50$	median	1.4998	3.7×10^{-4}	3.1516	23.374
	lower	1.497	2.5×10^{-4}	2.3246	10.223
	upper	1.502	5.27×10^{-4}	4.3794	79.794
	$\sqrt{\hat{R}}$	1.0002	1.0000	1.0003	1.0023
$\lambda^A = 25$	median	1.5008	3.61×10^{-4}	2.7186	15.159
	lower	1.499	2.53×10^{-4}	2.0737	7.954
	upper	1.503	4.99×10^{-4}	3.6074	36.871
	$\sqrt{\hat{R}}$	1.0000	1.0000	1.0054	1.0182
$\lambda^A = 10$	median	1.4993	2.54×10^{-4}	2.4543	11.639
	lower	1.498	2.03×10^{-4}	1.8867	6.597
	upper	1.502	3.89×10^{-4}	3.1804	24.056
	$\sqrt{\hat{R}}$	1.0061	1.0000	1.0000	1.0001
$\lambda^A = 5$	median	1.4992	2.28×10^{-4}	1.9280	6.876
	lower	1.497	1.58×10^{-4}	1.5022	4.492
	upper	1.502	3.17×10^{-4}	2.5070	12.268
	$\sqrt{\hat{R}}$	1.0008	1.0000	1.0007	1.0006
$\lambda^A = 1$	median	1.5005	3.46×10^{-4}	0.02317	1.023
	lower	1.494	1.97×10^{-4}	-0.3654	0.694
	upper	1.507	6.22×10^{-4}	0.3469	1.415
	$\sqrt{\hat{R}}$	1.0026	1.0024	1.0006	1.0007

Table 10.2 *Posterior summaries for data generated according to five simulation models where lower and upper relate to the lower and upper bounds of the 95% confidence interval for the parameter.*

Y, is modeled as in (10.9). Because of the modular structure of our computational tools, it is not difficult to compose posterior sampling and mode finding algorithms in this more general setting. Using the notation of Section 10.3.2, the joint probability model factors as

$$p(X, Y, Z, \theta^A, \theta^{-A}) = p(Y|Z, \theta^{-A})p(Z|X, \theta^A)p(X|\theta^{-A})p(\theta^{-A})p(\theta^A),$$
$$(10.19)$$

where Y is the observed data, X is the ideal data, Z is the ideal data after absorption, and θ^{-A} are all model parameters not involved in the absorption line. The first factor on the right-hand side of (10.19) represents the effects of blurring, background, effective area of the instrument, and any other absorption components in the model; the second factor represents the

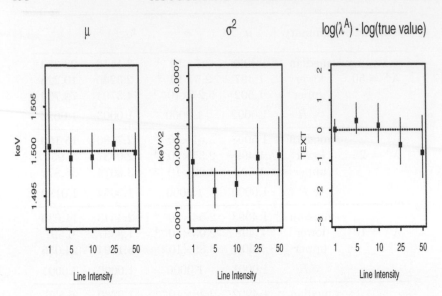

Figure 10.7 *Medians and 95% credible intervals for the five simulations. The horizontal axes represent the five simulations. Intervals in the third graph are translated (by substracting by the logarithm of the true value) so as to be comparable. In each case the dashed lines are the true values.*

absorption line; the third factor is the distribution of the ideal data given in (10.10); and the last two factors are independent prior distributions. Thus, we can construct a two step Gibbs sampler, perhaps with Metropolis-Hastings approximations, to obtain a sample from $p(X, Z, \theta^{-A}, \theta^{A}|Y)$ as follows:

STEP 1: Draw $(\theta^{A}_{(t)}, \theta^{-A}_{(t)})$ from

$$p(\theta^{A}, \theta^{-A}|X, Y, Z) = p(\theta^{A}|X, Y, Z)p(\theta^{-A}|X, Y, Z)$$

STEP 2: Draw $(X_{(t)}, Z_{(t)})$ from

$$
\begin{aligned}
p(X, Z|Y, \theta^{A}, \theta^{-A}) &= p(X|Y, Z, \theta^{A}, \theta^{-A})p(Z|Y, \theta^{A}, \theta^{-A}) \\
&= p(X|Z, \theta^{A}, \theta^{-A})p(Z|Y, \theta^{A}, \theta^{-A}) \quad (10.20)
\end{aligned}
$$

The equalities follow from the factorization in (10.19). Because the draws of θ^{A} and of X are exactly as described in Section 10.3 while the draws of θ^{-A} and Z are given in van Dyk et al. (2001) we can easily implement this MCMC sampler. EM can be adapted using similar arguments.

10.4.2 An Example

To explore the effect of an absorption line on the spectral model of Section 10.2.2 we analyze the X-ray spectrum of Quasar 3S 0014+813 (Kuhr et al. 1981), using data observed with the ASCA instrument in 1993 (Elvis et al. 1994). We used all 512 of the instrument energy bins, except for the unreliable bins below ~ 0.5 keV or above ~ 10 keV. We model the continuum as a power law, i.e., $\lambda^C(\theta^C, E_j) = \theta_1^C E_j^{-\theta_2^C}$, and add an inter-stellar absorption component, $\alpha_1(\theta_1^A, E) = \exp(-\theta_1^A/E)$. We account for background by setting θ_l^B equal to the (rescaled) counts in the corresponding bin of the background observation. Because the data are relatively informative for θ^C and θ_1^A, we use flat prior distributions on a variance stabilizing transformation of each parameter. A sample from the posterior distribution was obtained by running three MCMC chains according to the algorithm described in van Dyk et al. (2001) for 1000 iterations each. (We use the nesting methods described by van Dyk and Meng (2000) to reduce the autocorrelation in the resulting chains and produce three draws per iteration.) We then combined the last 2000 draws of each chain to form a sample of 6000 draws. The marginal posterior distributions of θ_1^C, θ_2^C and θ_1^A are represented by the shaded histograms in Figure 10.8.

To explore the effect of an absorption line on our analysis, we manually subtracted counts from nine adjacent bins near 1 keV to simulate an absorption line:

original counts	30	30	28	24	31	37	28	26	29
altered counts	15	10	5	0	0	0	5	10	15

Using the altered data we refit the model exactly as described above (not accounting for the absorption line), yielding the marginal posterior distributions depicted by the histogram with dashed lines in Figure 10.8. The presence of the unaccounted for absorption feature has both increased the posterior variance of all three parameters and has shifted the distributions away from their "true" values. Clearly inference based on this posterior is biased by the non-ignorable missing data caused by the absorption line. Thus, we refit the altered data, this time accounting for the absorption line component as described in Section 10.4. (We use a gamma prior on λ^A with $E(\lambda^A) = \text{var}(\lambda^A) = 2$.) The new marginal posterior (solid lines in Figure 10.8) match the original marginal posterior distributions closely; the bias caused by the absorption line has been removed.

Figure 10.9 shows the cumulative probability of membership of four clusters as a function of energy. The clusters correspond to the photons observed by the detector, photons lost to instrument response, photons absorbed by the inter-stellar media, and photons absorbed in the absorption line. Although these clusters are not specifically spatial in character, if the source were diffuse, we might expect their relative density to vary

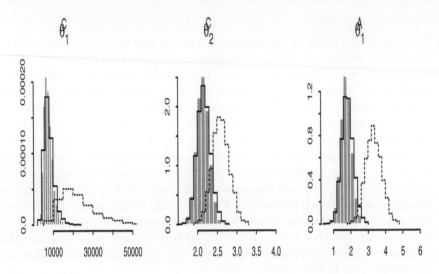

Figure 10.8 *Marginal posterior distributions. The shaded histograms represent the "true" model with no absorption line present. The dashed lines show the bias that an absorption line introduces to the parameter estimates, and the solid lines show the posterior distributions after the absorption line has been accounted for.*

across the source. Three of these clusters are completely unobserved—they are clusters within the idealized data, Y. Similar computations can separate the background, continuum, and emission line photon clusters, all of which are at least partially observed. If we confine our attention to the observed photons, the background, continuum, and emission line clusters are all sub-clusters of cluster 'A' in Figure 10.9. We use the posterior means of the model parameters to produce Figure 10.9; posterior variability can easily be used to compute error bars for the cluster probabilities.

10.5 Discussion

The statistical and computational challenges of image analysis in high energy astrophysics are truly immense. Accounting for the spatial, spectral, and temporal structure in the data along with the complexities in the photon redistribution matrix, pile-up, background contamination, and photon absorption requires highly structured models and sophisticated computing. Current work focuses on incorporating these complexities one at a time, taking advantage of the modular structure in both our models and computational techniques. The preliminary methods are useful for special classes

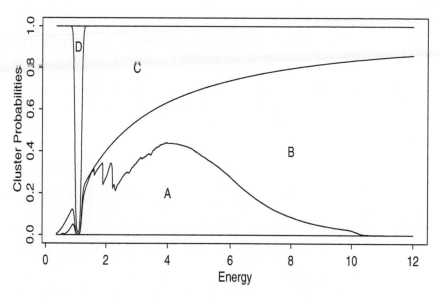

Figure 10.9 *Cumulative Cluster Probabilities. The figure displays the cumulative probabilities of four photon clusters as a function of Energy. Cluster 'A' contains observed photons; 'B' contains photons lost to instrumental response; 'C' contains photons absorbed by the smooth absorption term, α_1, e.g., due to the interstellar media, and 'D' contains photons absorbed in the absorption line. The relative size of the clusters vary dramatically as a function of Energy (the probabilities are computed using the posterior mean of the model parameters).*

of images (e.g., the spectral models can be applied directly to point sources) but need to be extended to be useful for more sophisticated images. Combining the spectral and spatial models is a particular area of active work.

Even within the much less complicated problem of accounting for absorption lines, there are sophisticated statistical and computational challenges. Handling the highly multi-modal posterior distribution will only become more complex as the overall model incorporates more of the features of the source and data collection mechanism. In some cases, the choice of prior distribution can be quite important and careful prior specification along with sensitivity analysis is required. In general, however, we expect the three steps of first exploring the posterior distribution with mode finding algorithms, second fitting the model via MCMC, and finally checking the model using posterior predictive checks to be a practical strategy. Thus far, we have found that directly modeling the stochastic features in the underlying images and data collection to be both powerful statistically and tractable computationally, and thus, a fruitful strategy for image reconstruction.

Acknowledgements

This work is a product of the collaborative effort of the Astro-Statistics Group at Harvard University whose members include A. Connors, D. Esch, P. Freeman, H. Kang, V. L. Kashyap, R. Protassov, D. Rubin, A. Siemiginowska, E. Sourlas , and Y. Yu. The authors gratefully acknowledge funding for this project partially provided by NSF grant DMS-01-04129 and by NASA Contract NAS8-39073 (CXC).

CHAPTER 11

Spatial Modelling of Count Data: A Case Study in Modelling Breeding Bird Survey Data on Large Spatial Domains

C.K. Wikle

11.1 Introduction

The North American Breeding Bird Survey (BBS) is conducted each breeding season by volunteer observers (e.g. Robbins et al. 1986). The observers count the number of various species of birds along specified routes. The collected data are used for several purposes, including the study of the range of bird species, and the variation of the range and abundance over time (Link and Sauer 1998). Such studies usually require spatial maps of relative abundance. Traditional methods for producing such maps are somewhat *ad hoc* (e.g., inverse distance methods) and do not always account for the special discrete, positive nature of the count data (e.g. Sauer et al. 1995). In addition, corresponding prediction uncertainties for maps produced in this fashion are not typically available. Providing such uncertainties is critical as the prediction maps are often used as "data" in other studies and for the design of auxiliary sampling plans.

We consider the BBS modeling problem from a hierarchical perspective, modeling the count data as Poisson, conditional on a spatially varying intensity process. The intensities are then assumed to follow a log-normal distribution with fixed effects and with spatial and non-spatial random effects. Model-based geostatistical methods for generalized linear mixed models (GLMMs) of this type have been available since the seminal work of Diggle et al. (1998). However, implementation is problematic when there are large data sets and prediction is desired over large domains. We show that by utilizing spectral representations of the spatial random effects process, Bayesian spatial prediction can easily be carried out on very large data sets over extensive prediction domains. General discussion of the role of such Bayesian hierachical random effect modelling is given in 1, and approaches to spatio-temporal modelling are found here in 12.

The BBS sampling unit is a roadside route 39.2 km in length. Over each route, an observer makes 50 stops, at which birds are counted by sight and sound for a period of 3 minutes. Over 4000 routes have been included in the North American survey, but not all routes are available each year. As might be expected due to the subjectivity involved in counting birds by sight and sound, and the relative experience and expertise of the volunteer observers, there is substantial observer error in the BBS survey (e.g. Sauer et al. 1994).

In this study, we are concerned with the relative abundance of the House Finch (*Carpodacus mexicanus*). Figure 11.1 shows the location of the sampling route midpoints and observed counts over the continental United States (U.S.) for the 1999 House Finch BBS. The size of the circle radius is proportional to the number of birds observed at each site. This figure suggests that the House Finch is more prevalent in the Eastern and Western U.S. than in the interior. Indeed, this species is native to the Western U.S. and Mexico. The Eastern population is a result of a 1940 release of caged birds in New York. The birds were being sold illegally in New York City as "Hollywood Finches" and were supposedly released by dealers in an attempt to avoid prosecution. Within three years there were reports of the birds breeding in the New York area. Because the birds are prolific breeders and their juveniles disperse over long distances, the House Finch quickly expanded to the west (Elliott and Arbib 1953). Simultaneously, as the human population on the west coast expanded eastward (and correspondingly, changed the environment) the House Finch expanded eastward as well. By the late 1990s, the two populations met in the Central Plains of North America.

From Figure 11.1 it is clear that there are many regions of the U.S. that were not sampled in the 1999 House Finch BBS. Our interest here is to predict abundance over a relatively dense network of spatial locations, every quarter degree of latitude and longitude. The network of prediction grid locations includes 228 points in the longitudinal and 84 in the latitudinal direction, for a total of 19,152 prediction grid locations.

11.2 The Poisson Random Effects Model

Consider the model for the count process $y(\boldsymbol{x})$ given a spatially varying mean process $\lambda(\boldsymbol{x})$:

$$y(\boldsymbol{x})|\lambda(\boldsymbol{x}) \sim Poisson(\lambda(\boldsymbol{x})). \tag{11.1}$$

The log of the spatial mean process is given by:

$$\log(\lambda(\boldsymbol{x})) = \mu + z(\boldsymbol{x}) + \eta(\boldsymbol{x}), \tag{11.2}$$

where μ is a deterministic mean component, $z(\boldsymbol{x})$ is a spatially correlated random component, and $\eta(\boldsymbol{x})$ is an uncorrelated spatial random compo-

Figure 11.1 *Observation locations for 1999 BBS of House Finch (Carpodacus mexicanus). Radius and color are proportional to the observed counts.*

nent. In general, the fixed component μ might be related to spatially varying covariates (such as habitat) and could include "regression" terms. We will consider the simple constant mean formulation in this application. The correlated process, $z(\boldsymbol{x})$, is necessary in this application because we have substantial prior belief that the counts at "nearby" routes are correlated. From a scientific point of view, this is likely due (at least in part) to the fact that the birds are attracted to specific habitats, and we know that habitat is correlated in space. Typically, one can view the z-process as accounting for the effects of "unknown" covariates, since it induces spatial structure in the λ-process, and thus the observed counts. In that sense, maps of the z-process may be interesting and lead to greater understanding as to the preferred habitat of the modeled bird species (e.g. Royle et al. 2001). The random component $\eta(\boldsymbol{x})$ accounts for observer effects. A major concern in the analysis of BBS data is the known observer bias, as discussed previously. Typically, we can assume that since the observers produce counts on different routes, they are independent with regard to space.

The above discussion suggests that we might model $z(\boldsymbol{x})$ as a Gaussian random field with zero mean and covariance given by $c_\theta(\boldsymbol{x}, \boldsymbol{x}')$, where θ represents parameters (possibly vector-valued) of the covariance function c. In addition, we assume $\eta(\boldsymbol{x}) \sim N(0, \sigma_\eta^2)$, where $\operatorname{cov}(\eta(\boldsymbol{x}), \eta(\boldsymbol{x}')) = 0$ if $\boldsymbol{x} \neq \boldsymbol{x}'$.

As presented, the Poisson spatial model follows the framework for generalized geostatistical prediction formulated in Diggle et al. (1998). An example of this approach applied to the BBS problem can be found in Royle et al. (2001). However, implementation in that case was concerned with relatively small data sets and over limited geographical regions. The Gaussian random field-based Bayesian hierarchical approach becomes increasingly difficult to implement as the dimensionality of the data and number of prediction locations increases. Consequently, such an approach is not feasible at the continental scale and high resolution that we require in the present application. However, as outlined in Royle and Wikle (2001), one can still use the Bayesian GLMM methodology in these high-dimensional settings if one makes use of spectral representations. This approach is summarized in the next section.

11.2.1 Spectral Formulation

Let $\{x_i\}_{i=1}^m$ be the set of data locations, at which counts $y(x_i)$ were observed. Further, let $\{x_j\}_{j=1}^n$ be the set of prediction locations, which may, but need not, include some or all of the m data locations. We now rewrite the mean-process model (11.2):

$$\log(\lambda(x_i)) = \mu + k_i' z_n + \eta(x_i), \tag{11.3}$$

where z_n is an $n \times 1$ vector representation of z-process at the prediction locations, and the vector k_i relates the log-mean process at observation location x_i to one or more elements of the z-process at prediction locations (e.g. Wikle et al. 1998, Wikle et al. 2001). We then assume:

$$z_n = \Psi \alpha + \epsilon, \tag{11.4}$$

where Ψ is an $n \times p$ matrix, fixed and known, α is a $p \times 1$ vector of coefficients with $\alpha \sim N(0, \Sigma_\alpha)$, and $\epsilon \sim N(0, \sigma_\epsilon^2 I)$. We let Ψ consist of spectral basis functions $[\psi_{j,k}]_{j=1,k=1}^{n,p}$ that are orthogonal. That is, if $\psi_k \equiv [\psi_{1,k}, \ldots, \psi_{n,k}]'$ then $\psi_k' \psi_j = 0$ if $k \neq j$ and 1, otherwise. In this case, we say that α are spectral coefficients. From a hierarchical perspective, we can write:

$$z_n | \alpha, \sigma_\epsilon^2 \sim N(\Psi \alpha, \sigma_\epsilon^2 I) \tag{11.5}$$

and

$$\alpha | \Sigma_\alpha \sim N(0, \Sigma_\alpha). \tag{11.6}$$

In general, the covariance function for the α-process depends on some parameters θ; we denote this covariance by $\Sigma_\alpha(\theta)$. The modeling motivation for the hierarchy is apparent if we note that the random z-process can be written $z_n \sim N(0, \Sigma_z(\theta) + \sigma_\epsilon I)$, where σ_ϵ^2 accounts for the "nugget effect" due to small scale variability. Given (11.4), the covariance function for the z-process can be written, $\Sigma_z(\theta) = \Psi \Sigma_\alpha(\theta) \Psi'$.

In principle, any set of orthogonal spectral basis functions could be used for $\boldsymbol{\Psi}$. For example, one could use the leading variance modes of the co-variance matrix $\boldsymbol{\Sigma}_z$. Such modes are just the eigenvectors that diagonalize the spatial covariance matrix and thus are just principal components. These spatial principal components are known as Empirical Orthogonal Functions (EOFs) in the geostatistical literature (e.g. Obled and Creutin 1986, Wikle and Cressie 1999). Such a formulation is advantageous because it allows for non-stationary spatial correlation and dimension reduction ($p << n$). Another possibility would be to use Fourier basis functions in $\boldsymbol{\Psi}$. This could apply if the prediction locations were defined in continuous space or on a grid. However, as we will demonstrate, if we choose a grid implementation, one need not actually form the matrix $\boldsymbol{\Psi}$, which would be problematic for grid sizes of order 10^5 as we consider here. That is, the operation $\boldsymbol{\Psi}\boldsymbol{\alpha}$ is actually an inverse Fourier transform operation on the vector $\boldsymbol{\alpha}$. On a discrete lattice, one can use Fast Fourier Transform (FFT) procedures to efficiently implement this transform without having to make or store the matrix of basis functions. In this case, $p = n$. If the z-process is stationary, the use of Fourier basis functions suggests that the matrix $\boldsymbol{\Sigma}_\alpha(\boldsymbol{\theta})$ is diagonal (asymptotically). For situations where it is more appropriate to assume that the process is nonstationary and the prediction locations can be thought of as a discrete grid, one could consider a wavelet basis function for $\boldsymbol{\Psi}$. In this case, the operation $\boldsymbol{\Psi}\boldsymbol{\alpha}$ is just an inverse discrete wavelet transform of $\boldsymbol{\alpha}$; again, $\boldsymbol{\Psi}$ need not be constructed directly. Depending on the class of wavelets chosen, the matrix $\boldsymbol{\Sigma}_\alpha(\boldsymbol{\theta})$ may be diagonal (asymptotically) or nearly so.

In the hierarchical implementation, the parameterization of $\boldsymbol{\Sigma}_\alpha(\boldsymbol{\theta})$ is especially critical. For example, with wavelet basis functions, we might assume a fractional scaling behavior in the variance of the different wavelet scales. This is particularly useful when the process is known to exhibit such behavior, such as turbulence examples in atmospheric science (e.g. Wikle et al. 2001). Alternatively, we might assume a common stationary class for the z-process, such as the Matérn class of covariance functions,

$$c(d_{ij}) = \phi(\theta_1 d_{ij})^{\theta_2} K_{\theta_2}(\theta_1 d_{ij}), \quad \phi > 0, \theta_1 > 0, \theta_2 > 0, \qquad (11.7)$$

where d_{ij} is the distance between two spatial locations, K_{θ_2} is the modified Bessel function, θ_2 is related to the degree of smoothness of the spatial process, θ_1 is related to the correlation range, and ϕ is proportional to the variance of the process (e.g. page 48, Stein 1999). The corresponding spatial spectral density function at frequency ω is,

$$f(\omega; \theta_1, \theta_2, \phi, g) = \frac{2^{\theta_2-1}\phi\Gamma(\theta_2 + g/2)\theta_1^{2\theta_2}}{\pi^{g/2}(\theta_1^2 + \omega^2)^{\theta_2+g/2}}, \qquad (11.8)$$

where g is the dimension of the spatial process (e.g. page 49, Stein 1999). Thus, if one chooses Fourier basis functions for $\boldsymbol{\Psi}$ and assumes the Matérn class, then $\boldsymbol{\Sigma}_\alpha(\boldsymbol{\theta})$ should be diagonal (asymptotically) with diagonal elements corresponding to f given by (11.8). If not known, one must specify prior distributions for $\boldsymbol{\theta}$ and ϕ at the next level of the model hierarchy.

11.2.2 Model Implementation and Prediction

The hierarchical Poisson model with a spectral spatial component is summarized as follows. The joint likelihood for all observations $\boldsymbol{y}$ (an $m \times 1$ vector) is

$$[\boldsymbol{y}|\boldsymbol{\lambda}] = \prod_{i=1}^{m} Poisson(\lambda(\boldsymbol{x}_i)), \qquad (11.9)$$

where $\boldsymbol{\lambda}$ is an $m \times 1$ vector, corresponding to the locations of the vector $\boldsymbol{y}$. The joint prior distribution for $log(\lambda(\boldsymbol{x}_i))$ is:

$$[\log(\boldsymbol{\lambda})|\mu, \gamma, \boldsymbol{z}_n, \sigma_n^2] = N(\mu\mathbf{1} + \gamma\boldsymbol{K}\boldsymbol{z}_n, \sigma_\eta^2\boldsymbol{I}), \qquad (11.10)$$

where $\mathbf{1}$ is an $m \times 1$ vector of ones, $\log(\boldsymbol{\lambda})$ is the $m \times 1$ vector with elements $\log(\lambda(\boldsymbol{x}_i))$, $\boldsymbol{K}$ is an $m \times n$ matrix with rows $\boldsymbol{k}_i'$, and γ is a scaling coefficient (introduced for computational reasons as discussed below). Then, let

$$[\boldsymbol{z}_n|\boldsymbol{\alpha}, \sigma_e^2] = N(\boldsymbol{\Psi}\boldsymbol{\alpha}, \sigma_e^2\boldsymbol{I}), \qquad (11.11)$$

and allow the spectral coefficients to have distribution,

$$[\boldsymbol{\alpha}|\boldsymbol{R}_\alpha(\theta_1)] = N(\mathbf{0}, \boldsymbol{R}_\alpha(\theta_1)), \qquad (11.12)$$

where $\boldsymbol{R}_\alpha(\theta_1)$ is a diagonal matrix. For the BBS illustration presented here, we let $\theta_2 = 1/2$ in (11.7) (i.e., we assume the covariance model is exponential) but assume the dependence parameter θ_1 is random. Note that as a consequence of including the γ parameter in (11.10) we are able to specify the conditional covariance of $\boldsymbol{\alpha}$ as the diagonalization of a correlation matrix rather than a covariance matrix (see discussion below). Finally, to complete the model hierarchy, we assume the remaining parameters are independent and specify the following prior distributions:

$$\mu \sim N(\mu_0, \sigma_\mu^2), \quad \sigma_\eta^2 \sim IG(q_\eta, r_\eta), \quad \gamma \sim U[0, b], \qquad (11.13)$$

$$\sigma_e^2 \sim IG(q_e, r_e), \quad \theta_1 \sim U[u_1, u_2], \qquad (11.14)$$

where $IG(\ )$ refers to an inverse gamma distribution, and $U[\ ]$ a uniform distribution. For the BBS House Finch data we select $q_\eta = 0.5$, $q_e = 1$, $r_\eta = 2$, $r_e = 10$, $\mu_0 = 0$, $\sigma_\mu^2 = 10$, $b = 100$, $u_1 = 1$, and $u_2 = 100$ (note, our parameterization of the exponential is $r(d) \propto \exp(-\theta_1 d)$, where d is the distance). These hyperparameters correspond to rather vague proper priors.

The alternative to specifying γ in (11.10) is to let the conditional covariance of $\boldsymbol{\alpha}$ be $\sigma_\alpha^2 \boldsymbol{R}_\alpha(\theta_1)$. However, as is often the case for Bayesian spatial models that are deep in the model hierarchy (and thus, relatively far from the data), the MCMC implementation has difficulty converging because of the tradeoff between the spatial process variance, σ_α^2, and the dependence parameter, θ_1. By allowing the z-process to have unit variance, as in the above formulation, we need not estimate σ_α^2 (which is 1 in this case). The variance in the spatial process is then achieved through γ. In situations where the implied assumption of homogeneous variance is unrealistic, a more complicated reparameterization would be required. Note that the γ parameterization also affects the interpretation of the variance of the z-process (i.e., $\sigma_e^2 = \sigma_\epsilon^2/\gamma^2$).

Our goal is the estimation of the joint posterior distribution,

$$[\log(\boldsymbol{\lambda}), \boldsymbol{z}_n, \theta_1, \gamma, \sigma_\eta^2, \sigma_e^2, \mu | \boldsymbol{y}] \quad \propto \quad [\boldsymbol{y}| \log(\boldsymbol{\lambda})][\log(\boldsymbol{\lambda})|\mu, \boldsymbol{z}_n, \sigma_\eta^2][\boldsymbol{z}_n|\boldsymbol{\alpha}, \sigma_e^2]$$
$$\times \quad [\boldsymbol{\alpha}|\theta_1][\theta_1][\gamma][\mu][\sigma_\eta^2][\sigma_e^2]$$

Although this distribution cannot be analyzed directly, we are able to use MCMC approaches as suggested by Diggle et al. (1998) to draw samples from this posterior and appropriate marginals. In particular, we utilized a Gibbs sampler with Metropolis-Hastings sampling of $\log(\boldsymbol{\lambda})$ and θ_1 (see Royle et al. 2001). Perhaps more importantly, we would like estimates from the posterior distribution of $\boldsymbol{\lambda}_n$, the λ-process at prediction grid locations. The key difficulty in the traditional (non-spectral) geostatistical formulation is the dimensionality of the full-conditional update for the z-process given all other parameters. As we show below, this is no longer a serious problem if we make use of the spectral representation.

Selected Full-Conditional Distributions

As mentioned above, for the most part the full-conditional distributions follow those outlined generally in Diggle et al. (1998) and specifically, those in Royle et al. (2001). However, the spectral representation allows simpler forms for the $\boldsymbol{z}_n$ and $\boldsymbol{\alpha}$ full-conditionals.

The full-conditional distribution for $\boldsymbol{z}_n$ can be shown to be:

$$\boldsymbol{z}_n|\cdot \sim N(\boldsymbol{S}_z^{-1}\boldsymbol{a}_z, \boldsymbol{S}_z^{-1}), \tag{11.15}$$

where $\boldsymbol{S}_z = \boldsymbol{I}/\sigma_e^2 + \boldsymbol{K}'\boldsymbol{K}\gamma^2/\sigma_\eta^2$ and $\boldsymbol{a}_z = \boldsymbol{\Psi}\boldsymbol{\alpha}/\sigma_e^2 + \boldsymbol{K}'(\log(\boldsymbol{\lambda}) - \mu\boldsymbol{1})\gamma/\sigma_\eta^2$. In our case, $\boldsymbol{K}$ is an incidence matrix (a matrix of ones and zeros) such that each observation is only associated with one prediction grid location (a reasonable assumption at the resolution presented here). Thus, $\boldsymbol{K}'\boldsymbol{K}$ can be shown to be a diagonal matrix with 1's and 0's along the diagonal. Although the matrix $\boldsymbol{S}_z$ is very high-dimensional (order $10^5 \times 10^5$), it is diagonal and trivial to invert. In addition, $\boldsymbol{\Psi}\boldsymbol{\alpha}$ can be calculated by the inverse FFT function (a fast operation) and $\boldsymbol{z}_n$ is updated as simple univariate normal distributions. In practice, we update these simultaneously in a matrix language implementation.

Similarly,

$$\boldsymbol{\alpha}|\cdot \sim N(\boldsymbol{S}_\alpha^{-1}\boldsymbol{a}_\alpha, \boldsymbol{S}_\alpha^{-1}), \tag{11.16}$$

where $S_\alpha = (\Psi'\Psi/\sigma_e^2 + R_\alpha(\theta_1)^{-1})$ and $a_\alpha = \Psi'z_n/\sigma_e^2$. At first glance, this appears problematic due to the $\Psi'\Psi$ and $R_\alpha(\theta_1)^{-1}$ terms in the full-conditional variance. However, since the spectral operators are orthogonal, $\Psi'\Psi = I$ and the matrix $R_\alpha(\theta_1)^{-1}$ is diagonal as discussed previously. Furthermore, $\Psi'z_n$ is just the FFT operation on z_n and is very fast. Thus,

$$\alpha|\cdot \sim N((I/\sigma_e^2 + R_\alpha(\theta_1)^{-1})^{-1}\Psi'z_n/\sigma_e^2, (I/\sigma_e^2 + R_\alpha(\theta_1)^{-1})^{-1}) \quad (11.17)$$

and can be sampled as individual univariate normals, or easily in a block update.

Prediction

To obtain predictions of λ_n, the λ-process at the prediction grid locations, we sample from

$$[\log(\lambda_n^{(t)})|z_n^{(t)}, \gamma^{(t)}, \mu^{(t)}, \sigma_\eta^{2\,(t)}] = N(\mu^{(t)}1 + \gamma^{(t)}z_n^{(t)}, \sigma_\eta^{2\,(t)}I), \quad (11.18)$$

where 1 is $n \times 1$ and $\mu^{(t)}$, $\gamma^{(t)}$, $z_n^{(t)}$, $\sigma_\eta^{2\,(t)}$ are the t-th samples from the MCMC simulation. We obtain $\lambda_n^{(t)}$ by simply exponentiating these samples.

Implementation

The MCMC simulation must be run long enough to achieve precise estimation of model parameters and predictions. For the BBS House Finch data, the MCMC simulation was run for 200,000 iterations after a 50,000 burn-in period. For sake of comparison, the algorithm took approximately 0.5 seconds per iteration with a MATLAB implementation on a 500 MHz Pentium III processor running Linux. Considering there are nearly 20,000 prediction locations and relatively strong spatial structure, this is quite fast. We examined many shorter runs to establish burn-in time and to evaluate model sensitivity to the fixed parameters and starting values. The model does not seem overly sensitive to these parameters.

11.3 Results

The posterior mean and posterior standard deviation for the scalar parameters are shown in Table 11.1.

Figure 11.2 shows the posterior mean for the gridded z-process. We note the agreement with the data shown in Figure 11.1. One might examine this map to indentify possible habitat covariates that are represented by the spatial random field. One possibility in this case might be elevation and population, both of which are thought to be associated with the prevalence of the House Finch.

We note that the prediction grid extends beyond the continental United States. Clearly, estimates over ocean regions are meaningless with regard to House Finch data. These estimates are a result of the large-scale Fourier

Table 11.1 *Posterior mean and standard deviation of univariate model parameters.*

Parameter	Posterior Mean	Posterior Standard Deviation
μ	0.74	0.105
γ	1.41	0.138
σ_η^2	0.84	0.100
σ_e^2	0.23	0.064
θ_1	14.78	4.605

coefficients in the model. Fortunately, the map of posterior standard deviations for this process, shown in Figure 11.3, indicates that these regions with no-data are highly suspect. This is also true of the northern plains region, which has few observations. Of course, having the prediction grid extend over the ocean is not ideal in this case, but the FFT-based algorithm requires rectangular grids. We could control for the land-sea effect by having an indicator covariate or possibly, a regime-specific model. Such modifications would be simple to implement in the hierarchical Bayesian framework presented here. However, simulation studies have shown that these are not necessary and if desired, one could simply mask the water portions of the map for presentation.

Finally, Figure 11.4 and Figure 11.5 show the posterior mean and standard deviation of the λ-process on the prediction grid. These plots show clearly that the posterior standard deviation is proportional to the predicted mean, as expected with Poisson count data. In addition, the standard errors are also high in data sparse regions, as we expect.

11.4 Conclusion

In summary, we have demonstrated how the Bayesian implementation of geostatistical-based GLMM Poisson spatial models can be implemented in problems with very large numbers of prediction locations. By utilizing relatively simple spectral transforms and associated orthogonality and decorrelation, we are able to implement the modeling approach very efficiently in general MCMC algorithms.

Acknowledgement

This research has been supported by a grant from the U.S. Environmental Protection Agency's Science to Achieve Results (STAR) program, Assis-

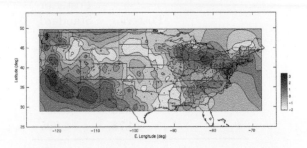

Figure 11.2 *Posterior mean of z_n for the 1999 BBS House Finch data.*

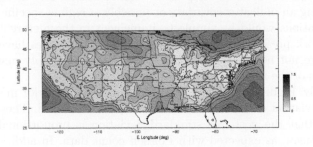

Figure 11.3 *Posterior standard deviation of z_n for the 1999 BBS House Finch data.*

tance Agreement No. R827257-01-0. The author would like to thank Andy Royle for providing the BBS data and for helpful discussions.

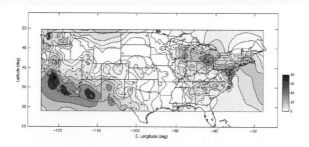

Figure 11.4 *Posterior mean of gridded* λ_n *for the 1999 BBS House Finch data.*

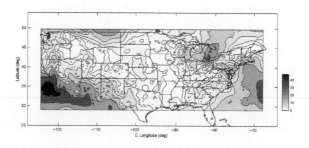

Figure 11.5 *Posterior standard deviation of* λ_n *for the 1999 BBS House Finch data.*

PART III

Spatio-temporal cluster modelling

CHAPTER 12

Modelling Strategies for Spatial-Temporal Data

J.T. Kent   *K.V. Mardia*

12.1 Introduction

Spatial-temporal modelling has largely been developed through applications in geostatistics, hydrology and meteorology. More recent activities in the area include environment monitoring, tracking, functional MRI, health data and facial analysis. For a recent snapshot of activities, see Mardia et al. (1999). Motivated by these applications, the field has adopted various modelling strategies. Current thinking in the field has been surveyed by Haslett (1989), Goodall and Mardia (1994), Kyriakidis and Journel (1999), Mardia et al. (1998), Wikle and Cressie (1999) and Brown et al. (2000).

The ideas behind these spatial-temporal models can be broadly cross-classified according to

(a) their motivation,

(b) their underlying objectives and

(c) the scale of data.

Under (a) the motivations for models can be classified into four classes: (i) extensions of time series methods to space (ii) extension of random field and imaging techniques to time (iii) interaction of time and space methods and (iv) physical models. Under (b) the main objectives can be viewed as either data reduction or prediction. Finally, under (c) the available data might be sparse or dense in time or space respectively, and the modelling approach often takes this scale of data into account. In addition, the data can be either continuously indexed or discretely indexed in space and/or time. Based on these considerations, especially (i) – (iii), we will describe several modelling strategies in detail and discuss their implementation. In Chapter 1 in this volume, space-time modelling concepts are introduced, while in Chapter 14, space-time modelling with a focus on object recognition is presented.

For simplicity we assume that the data take the form of a regular array in space-time. That is, we have one-dimensional observations

$$y_{ij}, \quad i = 1, \ldots, n, \ j = 1, \ldots, m, \tag{12.1}$$

at sites $\boldsymbol{x}_i \in \mathbb{R}^d$ and at times $t_j \in \mathbb{R}$. Typically, $d = 1, 2$, or 3 for observations on the line, in the plane or in 3D space. However, we do not usually assume that the sites are regularly-spaced in $\mathbb{R}^d$.

The objective is to model the data as

$$y_{ij} = z(\boldsymbol{x}_i, t_j) + \epsilon_{ij}, \quad \epsilon_{ij} \sim N(0, \sigma_0^2), \quad \sigma_0^2 \geq 0, \tag{12.2}$$

where $\{z(\boldsymbol{x}, t), \ \boldsymbol{x} \in \mathbb{R}^d, \ t \in \mathbb{R}\}$ is a stochastic or deterministic space-time process. We usually assume the error terms ϵ_{ij} are independent, identically distributed. In the language of geostatistics, such error terms are often known as a "nugget effect". Once a smooth process $\hat{z}(\boldsymbol{x}, t)$ has been fitted, it can be used for interpolation and prediction. Both sites and times may be either continuously or discretely indexed according to context.

Finally, we comment on notation. In general we use boldface to indicate vectors, e.g. $\boldsymbol{\gamma}$. In particular, sites in $\mathbb{R}^d$ will be denoted by $\boldsymbol{x}$ with components $(x[1], \ldots, x[d])$.

12.2 Modelling Strategy

A key property of much spatial-temporal data is spatial-temporal continuity; that is, observations at nearby sites and times will tend to be similar to one another. This underlying smoothness of a process $z(\boldsymbol{x}, t)$ can be captured in the following ways:

- *parametrically*, using a finite-dimensional space of regression or drift functions, or

- *nonparametrically*, using autocorrelation to make nearby values similar.

Both of these approaches can be applied in space and/or time. Letting D and C stand for a "drift" and "correlation" approach, respectively, the following types of models can be considered:

(a) (D-D) Tensor products of drift in time and drift in space. This approach is explored in Section 3.

(b) (D-C) Drift in space and correlation in time. The Kriged Kalman filter model of Mardia et al. (1998) and Wikle and Cressie (1999) exemplifies this approach; see Section 4.

(c) (C-D) Correlation in space and drift in time. Not explored here.

(d) (C-C) Joint correlation in space and time. Examples of this approach include space-time autoregressive and related models (Section 6.8 Cressie 1993) in discrete space-time and the "diffusion-injection" model in continuous space-time (pp. 430–433 Whittle 1986).

The density of data points in space and time can guide the choice of modelling strategy. When the data are sparse, there is often a preference for drift-style models, as there is not enough information to fit an autocorrelation structure. Of course a disadvantage of regression models is that the class of fitted curves and surfaces can be rather inflexible, especially for prediction and extrapolation.

On the other hand, when the data are rich, autocorrelation models become more feasible and flexible. In particular, they allow for more adaptive prediction and extrapolation. It is well-known that there is a close link between the use of autocorrelation models and the use of splines to fit curves and surfaces to discrete data; see, e.g., Kent and Mardia (1994), Chapter 3 of Wahba (1990) and pp. 180–183 of Cressie (1993). Thus, the use of autocorrelation models has a nonparametric flavour to it.

12.3 D-D (Drift-Drift) Models

In this section we consider models which involve drift functions in space $x \in \mathbb{R}^d$ and time $t \in \mathbb{R}$. There are three ingredients to these models:

(a) $\mathcal{F}$: a p-dimensional vector space of functions $\mathbb{R}^d \to \mathbb{R}$, specifying how the random field can vary spatially. Let $\{f_{(\alpha)}(x) : \alpha = 1, \ldots p\}$ denote a basis.

(b) $\mathcal{G}$: a q-dimensional vector space of functions $\mathbb{R} \to \mathbb{R}$ specifying how the random field can vary temporally. Let $\{g_{(\beta)}(t) : \beta = 1, \ldots q\}$ denote a basis.

(c) r : a rank, $r > 0$ representing complexity of the model.

Both $\mathcal{F}$ and $\mathcal{G}$ will usually include the constant function to accommodate an intercept term. From these ingredients we form a deterministic spatial-temporal process of the form

$$z(x, t) = \sum_{\alpha=1}^{p} \sum_{\beta=1}^{q} a_{\alpha\beta} f_{(\alpha)}(x) g_{(\beta)}(t), \tag{12.3}$$

where the $p \times q$ matrix of coefficients $A = (a_{\alpha\beta})$ has rank r.

Let the matrix $F(n \times p)$ denote the values of the spatial basis functions at the sites x_i, $i = 1, \ldots, n$. Similarly, let the matrix $G(m \times q)$ denote the values of the temporal basis functions at the times t_j, $j = 1, \ldots, m$. Thus, the model (12.2) takes the form

$$Y = F A G^T + E, \tag{12.4}$$

where $Y = (y_{ij})$ and $E = (\epsilon_{ij})$. If the basis functions are chosen so that F and G have orthonormal columns, then A can be estimated using the dominant r components in a singular value decomposition of $F^T Y G$ (Kent et al. 2001), though if an intercept term is separated out, estimation can be a bit more involved.

In this section we shall explore various choices for the space of drift functions and give an example.

Choices of Drift Functions

(a) Polynomials. Low order polynomials form the simplest choice. They do not depend on the choice of sites (for $\mathcal{F}$) or on the choice of times (for $\mathcal{G}$). A disadvantage, especially for high order, is that they are rather wild in their oscillatory behavior and they grow rapidly as $|x| \to \infty$ or $t \to \infty$, which can lead to unrealistic extrapolations and predictions.

(b) Trigonometric functions. Fourier expansions can be very useful for data which are periodic in space.

(c) Principal kriging functions. These are a set of functions in space or time adapted to the locations of the sites or times, respectively, and are arranged from coarse scale to fine scale in terms of variation.

Next we describe how to constuct the principal kriging functions. For simplicity we largely focus on functions of time. There are two main ingredients in the construction. First is a vector space of functions $\mathcal{G}_0$ of dimension $q_0 \geq 0$ called the "null space", and which will form a subspace of $\mathcal{G}$. Second is a "potential" function $\tau(t)$, say, which is conditionally positive definite with respect to $\mathcal{G}_0$. That is, for all distinct times $t^{(k)}$, $k = 1, \ldots, k_0$ and all vectors of coefficients $\boldsymbol{\delta} = (\delta_k, \ k = 1, \ldots, k_0) \neq 0$,

$$\sum_{k_1, k_2 = 1}^{k_0} \delta_{k_1} \delta_{k_2} \tau(t^{(k_1)} - t^{(k_2)}) > 0$$

whenever $\sum_{k=1}^{k_0} \delta_k g(t^{(k)}) = 0$ for all $g \in \mathcal{G}_0$.

Given the data times t_j, $j = 1, \ldots, m$, define $T(m \times m)$ by

$$T = (\tau(t_i - t_j)).$$

Next define an $m \times q_0$ "drift" matrix U by $u_{j\beta} = g_{0\beta}(t_j)$, where $\{g_{0\beta}(t), \ \beta = 1, \ldots, q_0\}$ is a basis of functions in $\mathcal{G}_0$. It is assumed that the data times are suitably spaced so that this matrix is of full rank q_0. These matrices can be combined into an $(m + q_0) \times (m + q_0)$ matrix

$$K = \begin{bmatrix} T & U \\ U^T & 0 \end{bmatrix}.$$

with partitioned inverse

$$K^{-1} = \begin{bmatrix} B & A \\ A^T & C \end{bmatrix}, \quad \text{say,}$$

which, in particular, defines the matrices B and A. Denote the eigenvectors of B by $\boldsymbol{\gamma}_k$, $k = 1, \ldots, n$, with the eigenvalues written in nondecreasing order. The first q_0 eigenvalues of B are 0 and the corresponding eigenvectors are given by the q_0 columns of U. Finally, let $\boldsymbol{\tau}_0(t)$ denote the vector function of t with jth element $\tau_0(t)_j = \tau(t - t_j)$, $j = 1, \ldots, m$, and let $\boldsymbol{u}_0(t)$ denote the vector function of t with βth component $g_{0\beta}(t)$, $\beta = 1, \ldots, q_0$.

The kth principal kriging function is defined by

$$g_k^{(PKF)}(t) = \gamma_k^T (B\tau_0 + Au_0(t)).$$

It turns out that the principal kriging functions depend only on the span of the columns of U, not on the set of basis functions used to construct U. Also, it can be shown that $g_k^{(PKF)}(t)$ is an interpolating function with $g_k(t_j) = (\gamma_k)_j$. The vector space $\mathcal{G}$ is defined to be the span of $\{g_\beta^{(PKF)}(t) : 1 \leq \beta \leq q\}$, where q is a specified dimension, $q_0 \leq q \leq m$. For further details see, for example, Mardia et al. (1996).

One possible common choice for $\tau(t)$ is any valid covariance function for a stationary stochastic process in time, for which any null space of functions $\mathcal{G}_0$ will suffice. Another possible choice is $\tau(t) = |t|^3$, which is conditionally positive definite whenever $\mathcal{G}_0$ contains the linear functions, $\mathrm{span}(1, t)$. In this case it turns out that the kth principal kriging function is an interpolating cubic spline which minimizes the penalty functional

$$\Phi(g) = \int_{-\infty}^{\infty} \left(\frac{d^2 g(t)}{dt^2} \right)^2 dt.$$

subject to the constraints $g(t_j) = (\gamma_k)_j$, $j = 1, \ldots, m$. This example motivates the alternative name "principal spline" for a principal kriging function.

A plot of some principal splines for $n = 10$ equally spaced time points is given in Figure 12.1. Note how the functions appear to mimic the qualitative behavior of the successive polynomials in t. However, note that the principal splines grow only linearly as $|t| \to \infty$, more slowly than the quadratic and higher order polynomials.

Principal splines can also be constructed for two-dimensional data at sites x_i, $i = 1, \ldots, n$. In this case it is common to use the thin-plate spline penalty

$$\Phi(f) = \int \left\{ \left(\frac{\partial^2 f}{\partial x[1]^2} \right)^2 + 2 \left(\frac{\partial^2 f}{\partial x[1] \partial x[2]} \right)^2 + \left(\frac{\partial^2 f}{\partial x[2]^2} \right)^2 \right\} d\boldsymbol{x}$$

where the integral is over $\mathbb{R}^2$ and $\boldsymbol{x} = (x[1], x[2])$ denotes the components of $\boldsymbol{x}$. The above construction of principal splines can be carried out with little change in this setting as well. Replace the potential function by $\tau(\boldsymbol{x}) = |\boldsymbol{x}|^2 \log |\boldsymbol{x}|$, $\boldsymbol{x} \in \mathbb{R}^2$, and the null space by $\mathcal{G}_0 = \mathrm{span}(1, x[1], x[2])$.

Principal splines in $\mathcal{R}^2$ were introduced by Bookstein (1989) in the context of deformations in shape analysis, and he used the term "principal warps". The matrix B was called the "bending energy matrix".

Principal splines have some advantages over polynomials. First they grow less quickly than polynomials outside the domain of the data (linearly for cubic splines) and so are more stable for extrapolation. Second, they are

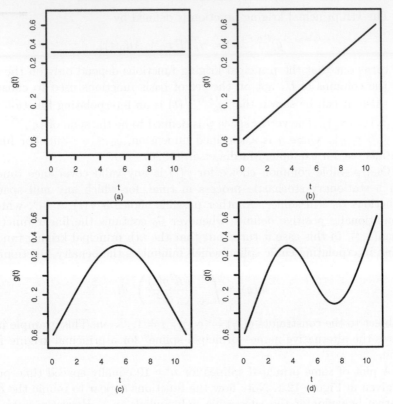

Figure 12.1 *First four principal cubic splines for $n = 10$ equally spaced time points: (a) and (b) are the constant and linear function corresponding to the null space $\mathcal{G}_0$; (c) and (d) are analogous to a quadratic and cubic function.*

adaptive to the given arrangement of times or sites, so that they can represent more detailed behavior where the times or sites are most dense.

Example – Growth in Rat Skulls

Chapter 7 of Bookstein (1991) analyzes a set of rat data on which 8 landmarks have been identified on a two-dimensional midsagittal section of the calvarium (the skull without the lower jaw). These measurements have been repeated at m=8 times on each of 18 rats. The purpose of the analysis is to model the growth of the calvarium. In this case $\mathcal{F}$, in a minor variation of the above framework, is given by a vector space of mappings from $\mathbb{R}^2$ to $\mathbb{R}^2$, to represent deformations. Before any further analysis, it is necessary to do a Procrustes registration of these configurations of landmarks in order to focus on shape changes. After this registration there are $n = 12$ degrees of freedom remaining in the 8 landmarks × 2 components = 16 total degrees of freedom.

In this example we took $\mathcal{F}$ to be a full 12-dimensional subspace generated by pairs of principal thin-plate splines (after adjusting for Procrustes registration) and $\mathcal{G}$ to be a full 7-dimensional subspace generated by the principal cubic splines (after removing 1 degree of freedom for a 12-dimensional intercept term). A good fit was obtained by taking the rank $r = 2$. A plot of the data and the fitted models with $r = 1$ and $r = 2$ are given in Figure 12.2. See Kent et al. (2000,(2001)) for further analysis.

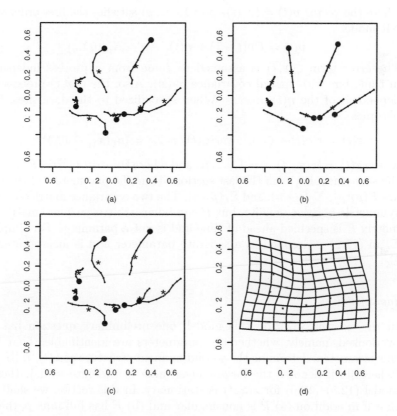

Figure 12.2 *Fitted growth models for the rat data: (a) raw data, averaged over individuals, (b) rank 1 model , (c) rank 2 model , all with the growth patterns blown up by a factor of 5 for clarity. Each "*" represents a landmark of the Procrustes mean shape, and each closed circle represents the position of a landmark at the initial time. Part (d) shows the grid deformation, without an expansion factor, between the initial and final times for the rank 2 model.*

12.4 D-C (Drift-Correlation) Models

In this section we modify (12.3) to get a process continuous in space and discrete in time. The vector space of spatial functions $\mathcal{F}$ remains unchanged, but for the time component, we replace the deterministic functions by a vector AR process, replacing the terms $\sum_\beta a_{\alpha\beta} g_{(\beta)}(t)$ by a stochastic process $v_\alpha(t)$ plus an error term. Thus,

$$z(\boldsymbol{x}, t) = \sum_{\alpha=1}^{p} f_{(\alpha)}(\boldsymbol{x}) v_\alpha(t) + \zeta(\boldsymbol{x}, t), \quad \boldsymbol{x} \in \mathbb{R}^d, \ t \in \mathbb{Z}, \qquad (12.5)$$

where the vector $\boldsymbol{v}(t) = (v_\alpha(t), \ \alpha = 1, \ldots, p)$ satisfies the first order vector AR model

$$\boldsymbol{v}(t) = P\boldsymbol{v}(t-1) + \boldsymbol{\eta}(t), \quad \boldsymbol{\eta} \sim N_p(0, \Sigma_\eta). \qquad (12.6)$$

The error term $\zeta(\boldsymbol{x}, t)$ is assumed to come from a process independent in time, but with spatial covariance $\sigma_\zeta(\boldsymbol{x})$. Also, the $\zeta(\boldsymbol{x}, t)$-process is independent of the $\boldsymbol{\eta}(t)$-process. When specialized to the data sites, (12.5) becomes

$$\boldsymbol{z}(t) = F\boldsymbol{v}(t) + \boldsymbol{\zeta}(t), \quad \text{var}(\boldsymbol{\zeta}(t)) = \Sigma_\zeta = (\sigma_\zeta(\boldsymbol{x}_{i_1} - \boldsymbol{x}_{i_2})), \qquad (12.7)$$

where $\boldsymbol{z}(t) = (z(\boldsymbol{x}_i, t), \ i = 1, \ldots, n)$ and where the drift matrix $F(n \times p)$ has the same meaning as in the last section. Thus the parameters of the model are $P(p \times p)$, $\Sigma_\zeta(n \times n)$, and $\Sigma_\eta(p \times p)$. The two covariance matrices must be symmetric positive definite, but P can be an arbitrary square matrix. The matrix F is specified ahead of time and is not a parameter. For simplicity, Σ_ϵ in (12.2) is dropped as a separate parameter and is incorporated into Σ_ζ.

Identifiability

In the existing work on this model, one preliminary question has been overlooked, namely, whether the parameters are identifiable. That is, do the parameters determine the second moment structure of $\boldsymbol{z}(t)$. If $||P|| < 1$ (where $|| \cdot ||$ denotes the largest eigenvalue in absolute value), then the model (12.5)–(12.6) for $z(\boldsymbol{x}, t)$ is stationary. In this section we shall show that if in addition (a) P is nonsingular and (b) F has full rank p, then the second moment structure of $\boldsymbol{z}(t)$,

$$A_h = E\{\boldsymbol{z}(t)\boldsymbol{z}^T(t-h)\}, \quad h \in \mathbb{Z}, \ h \geq 0,$$

is determined by the parameters.

First let $F = ULV^T$ be the singular value decomposition of F, where $U(n \times p)$ and $V(p \times p)$ are column orthonormal, and $L(p \times p)$ is diagonal with positive elements. Set $F^- = VL^{-1}U^T$ to be the Moore-Penrose generalized inverse of F, so that in particular $F^- F = I_p$. Set $\boldsymbol{z}'(t) = F^- \boldsymbol{z}(t)$, $\boldsymbol{\zeta}'(t) =$

$F^-\zeta(t)$, $\Sigma'_\zeta = F^- \, \Sigma_\zeta (F^-)^T$. Then

$$z'(t) = v(t) + \zeta'(t). \tag{12.8}$$

Define the matrix autocovariances

$$A'_h = E\{z'(t)z'^T(t-h)\}, \quad B_h = E\{v(t)v^T(t-h)\}, \quad h \ge 0. \tag{12.9}$$

A straightforward recursive expansion of (12.6) shows that

$$B_0 = \sum_{j=0}^{\infty} P^j \Sigma_\eta (P^T)^j, \quad B_h = P^h B_0, \quad h > 0, \tag{12.10}$$

and substituting (12.8) in (12.9) yields

$$A'_0 = B_0 + \Sigma_\zeta, \quad A'_h = B_h, \ h > 0. \tag{12.11}$$

Knowing the second moments $\{A_h, \ h \ge 0\}$ of the $z(t)$-process determines the second moments $\{A'_h, \ h \ge 0\}$ of the $z'(t)$-process. Provided P is non-singular, these quantities determine P, e.g. through

$$P = A'_2(A'_1)^{-1}. \tag{12.12}$$

Once P has been determined, B_0 can be found using $B_0 = P^{-1}A'_1$ from (12.10)–(12.11). Further, Σ_η can be found from P and B_0 through the identity $B_0 = \Sigma_\eta + PB_0 P^T$, which in turn can be derived from (12.10). Finally, Σ_ζ can be determined from the identity $A_0 = FB_0 F^T + \Sigma_\zeta$.

If P is singular, this approach breaks down. Indeed it is not possible to guarantee identifiablity in this case without some further restrictions on the parameters.

An Environmental Example

Mardia et al. (1998) analyzed $m = 707$ daily sulphur dioxide readings from March 29th 1983 to March 6th 1985. There are $n = 14$ monitored sites in Leeds and its vicinity; see Figure 12.3. This dataset is a part of a larger collection recorded in the UK from 1950s up to the present. The sites largely split into two clusters, an urban area (Leeds) and a rural area. During this period a clean air act was introduced. The rural area is a mining area which is close to a power station and the clean air act was not fully applied. One aim is to investigate whether there is any movement of pollution between the two areas.

The KKF model was implemented in Mardia et al. (1998) using principal kriging functions in space based on the covariance function

$$\tau(x) = e^{-(1.3)\sqrt{|x|}},$$

with a 3-dimensional null space given by the linear functions. It was found that using polynomials for the drift space did not fit the data as well as the principal kriging functions. Also, Σ_ζ was taken to be diagonal, which both simplifies the model and allows for some modest nonstationarity.

Various exploratory analyses were carried out using maximum likelihood estimation for some parameters. A small sequence of fitted surfaces is plotted in Figure 12.4. One conclusion which stands out from this figure is that on days 324 (15th Feb. 1984) and 325, there is a sudden jump in the overall pollution level. Curiously, over this period, highest levels are in Leeds and the lowest levels are in the mining area, the opposite of the typical spatial pattern at other times.

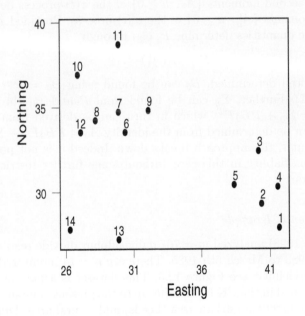

Figure 12.3 *Map of the sites for the environmental example. Leeds is in the upper left and the main mining areas are in the lower right.*

12.5 C-C (Correlation-Correlation) Models

In this section we develop a spatial-temporal model that is wholly stochastic. In discrete time and space, a spatial-temporal autoregressive model

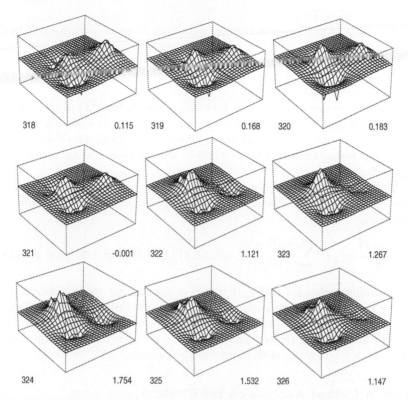

Figure 12.4 *Fitted surfaces for 9 time points 318, 319, ..., 326, based on 5 principal kriging functions. The time label appears at the lower left of each plot. At the lower right is a second label giving the distance of the mean of the surface at this time from the overall mean of the data.*

takes the form

$$z(\boldsymbol{x}, t+1) = \sum_{\boldsymbol{u}} w_{\boldsymbol{u}} z(\boldsymbol{x} + \boldsymbol{u}, t) + \zeta(\boldsymbol{x}, t), \quad \boldsymbol{x} \in \mathbb{Z}^d, \ t \in \mathbb{Z}, \qquad (12.13)$$

where $\boldsymbol{u}$ varies in a neighborhood of the origin including $\boldsymbol{u} = \boldsymbol{0}$, and the weights $w_{\boldsymbol{u}}$ represent a "smearing" or "blurring" effect, so that $z(\boldsymbol{x}, t+1)$ depends not just on $z(\boldsymbol{x}, t)$, but also on spatially neighboring values. The $\zeta(\boldsymbol{x}, t)$ are assumed to be a stationary process over space (often independent in this discrete setting) and independent for different times. Often the weights $w_{\boldsymbol{u}}$ are taken to be nonnegative. In this case a necessary condition for stationarity is $\sum_{\boldsymbol{u}} w_{\boldsymbol{u}} < 1$.

One possibility is for the weights to depend just on nearest neighbors, e.g. in two dimensions,

$$w_{\boldsymbol{0}} = \delta_1, \ w_{(\pm 1, \pm 1)} = \delta_2,$$

with $\delta_1 + 4\delta_2 < 1$.

A version of this process that is continuous in space and discrete in time is given by

$$z(\boldsymbol{x}, t+1) = \int z(\boldsymbol{x} - \boldsymbol{u})\phi(\boldsymbol{u})\, d\boldsymbol{u} - \psi^2 z(\boldsymbol{x}, t) + \zeta(\boldsymbol{x}, t) \qquad (12.14)$$

where $\phi(\boldsymbol{u})$ denotes the density function of the $N_d(\boldsymbol{0}, \lambda^2 I)$ distribution and for each integer t, $\zeta(\boldsymbol{x}, t)$ represents a process which is spatially stationary and independent for different times. Also, the parameter ψ^2 represents a damping term. Provided $0 < \psi^2 < 1$, the process will be stationary in space-time.

Letting time also be continuous leads to a continuous spatial-temporal process which can be described in terms of a stochastic partial differential equation

$$\frac{\partial z(\boldsymbol{x}, t)}{\partial t} = \frac{1}{2}\sigma^2 \sum_{l=1}^{d} \frac{\partial z^2(\boldsymbol{x}, t)}{\partial^2 x[l]} - \rho^2 z(\boldsymbol{x}, t) + \zeta(\boldsymbol{x}, t), \quad \boldsymbol{x} \in \mathbb{R}^d, \ t \in \mathbb{R}. \ (12.15)$$

The simplest choice for $\zeta(\boldsymbol{x}, t)$ is now a continuous space-time version of white noise, which makes sense only as a generalized random function. See, for example, Whittle (1986), Jones and Zhang (1997) and Brown et al. (2000) for this and related processes. It is necessary to take the damping term $\rho^2 > 0$ to ensure stationarity of the process.

12.6 A Unified Analysis on the Circle

Perhaps the simplest setting in which to compare these different modelling approaches is on a circle. Let $x \in [0, 2\pi)$ represent sites on the circle, and let time be continuous (for the D-D model) or discrete (for the D-C and C-C models). A natural set of functions for $\mathcal{F}$ is the set of sines and cosines up to order p', say, so that $p = 2p' + 1$ (one degree of freedom for the constant term $\cos(0 \cdot x) = 1$, and two degrees of freedom for each order $\alpha \geq 1$, $\cos\alpha x$ and $\sin\alpha x$).

Thus, the D-D model takes the form

$$z(x, t) = g_{c0}(t) + \sum_{\alpha=1}^{p'} \{g_{c\alpha}(t)\cos(\alpha x) + g_{s\alpha}(t)\sin(\alpha x)\}, \qquad (12.16)$$

where the coefficients $g_{c\alpha}(t)$, $g_{s\alpha}(t) \in \mathcal{G}$ are time-varying nonrandom Fourier coefficients.

The model D-C is similar, but the coefficients are now random processes and time is discrete. Thus,

$$z(x, t) = v_{c0}(t) + \sum_{\alpha=1}^{p'} \{v_{c\alpha}(t)\cos(\alpha x) + v_{s\alpha}(t)\sin(\alpha x)\} + \zeta(x, t), \quad (12.17)$$

where the $\{\zeta(x,t)\}$ are stationary on the circle and independent for different times.

Here the set of $\{v_{c\alpha}(t),\ v_{s\alpha}(t)\}$ follows a vector AR(1) process. If P and Σ_η are diagonal, these AR(1) processes are independent. Typically $v_{c0}(t)$ includes a nonzero mean to allow for an overall mean effect in the spatial-temporal process, but the other coefficients have mean 0. Further, since $\{\zeta(x,t)\}$ is stationary on the circle, it can be shown that if $z(x,t)$ is represented in a Fourier series, then its Fourier coefficients up to order p' follow independent ARMA(1,1) processes.

Last we turn to the C-C models. The periodic version of the continuous-space discrete-time model (12.14) can be given an infinite Fourier series representation

$$z(x,t) = z_{c0}(t) + \sum_{\alpha=1}^{\infty} \{z_{c\alpha}(t)\cos(\alpha x) + z_{s\alpha}(t)\sin(\alpha x)\}. \qquad (12.18)$$

The coefficients satisfy the independent AR(1) models,

$$z_{c0}(t+1) = (1-\psi^2)z_{c0}(t) + \zeta_{c0},$$
$$z_{c\alpha}(t+1) = \exp(-\frac{1}{2}\alpha^2\lambda)z_{c\alpha} + \zeta_{c\alpha}, \qquad (12.19)$$
$$z_{s\alpha}(t+1) = \exp(-\frac{1}{2}\alpha^2\lambda)z_{s\alpha} + \zeta_{s\alpha}, \quad \alpha \geq 1,$$

where $\zeta_{c\alpha}$, $\zeta_{s\alpha}$ are the Fourier coefficients of $\zeta(x,t)$, with independent $N(0,\sigma_\alpha^2)$ distributions. The variances σ_α^2 are the Fourier coefficients of the stationary covariance function on the circle of $\zeta(x,t)$ for each t. The exponential terms arise as Fourier coefficients of the wrapped normal distribution (page 50 Mardia and Jupp 2000). Thus, except for the range of summation, this model is a special case of (12.17) with P and Σ_η diagonal and $\Sigma_\zeta = 0$ in (12.17). As in that setting, a nonzero mean is usually included for z_{c0}. Note that the role of the error terms $\eta_{c\alpha}$, $\eta_{s\alpha}$ in the construction of the AR(1) processes $v_{c\alpha}$, $v_{s\alpha}$ below (12.17) using (12.6) is analogous to the role of the error terms $\zeta_{c\alpha}$, $\zeta_{s\alpha}$ to construct the AR(1) processes for $z_{c\alpha}$, $z_{s\alpha}$ in (12.19).

12.7 Discussion

We have tried to give a unified strategy to space-time modelling by distinguishing between drift and autocorrelation. More practical examples need to be studied to compare the merits of different approaches. However, there are also other considerations that can be used to guide a modelling strategy. Here is a brief summary of some of the issues.

1. Instead of our stochastic process approach to C-C models, it is possible

to specify directly the space-time autocovariance structures; e.g. Cressie and Huang (1999) and Glasbey (1998).

2. The simplest such models are separable in space and time; see, for example, Mardia and Goodall (1993) and Kyriakidis and Journel (1999).

3. Physical considerations, e.g. in meteorological data, can be used to guide model formulation (Wikle et al. 2001).

4. There have also been important advances in Bayesian hierarchical modelling; see, for example, Handcock and Wallis (1994), Wikle et al. (2001), and Chapter 11.

5. We have not discussed the subject of space-time point patterns. For a recent work see, for example, Smith and Robinson (1997) and other chapters in this volume, for example Chapter 14.

Acknowledgement

We wish to express our thanks to Sujit Sahu for helpful comments.

CHAPTER 13

Spatio-Temporal Partition Modelling: An Example from Neurophysiology

Peter Schlattmann   *Jürgen Gallinat*   *Dankmar Böhning*

13.1 Introduction

Neurophysiology tries to understand the function of the brain. Frequently activation experiments using positron emission tomography (PET) or magnetic resonance imaging (MRI) are undertaken to map the activation of certain areas of the brain following an experimental stimulus. Such functional mapping experiments produce data consisting of three-dimensional images where voxel values are indicative of regional neuronal activity. Usually no prior knowledge is available and thus the analysis proceeds at the voxel level.

In this setting images of statistics are formed where each voxel has an associated value of a simple statistic. These statistics express the activation at the voxel level. Many efforts have been undertaken in order to provide methods for the classification of individual voxels in neuro-imaging. More precisely this attempt deals with two problems at a time: For one to assess if there are any functional differences in an activation experiment and secondly to find methods which allow to address the activity of a voxel at a certain location.

One of the earliest attempts was the approach by Duffy et al. (1981) who apply z-scores or t-scores to EEG scalp data. Friston et al. (1990) use an ANOVA model and Holmes et al. (1996) use a randomization test in order to investigate these two questions. In the following we describe the application of methods from spatial epidemiology (Schlattmann and Böhning 1993) and meta analysis (Böhning et al. 1998, Böhning 1999) to the neuro-imaging problem. A review of spatial partition modelling is found in Chapter 7 in this volume.

13.2 The Neurophysiological Experiment

An interesting tool to investigate cerebral activity in the time range of milliseconds are auditory evoked potentials (AEPs) which occur after acoustic stimulation of an individual. One of the most prominent AEP is the N1-

component with a characteristic potential distribution over the head surface. From the physiological and pathophysiological point of view it is of special interest to locate the generators of this potential providing a more detailed knowledge about normal or disturbed brain function. Therefore, a method which locates the cerebral generators of the N1-component from the potential distribution on the surface of the head, is of theoretical and practical interest (Gallinat and Hegerl 1994).

The neurophysiological experiment comprised 22 healthy individuals who were repeatedly presented an auditory stimulus. Evoked responses were recorded with 32 electrodes referred to the central electrode labeled Cz. Subjects were seated with closed eyes in a slightly reclined chair with a head rest. An auditory oddball paradigm was employed with frequent non-targets (175 double clicks) and rare targets (55 tones, 1000 Hz). Tones (83 dB SPL, 40 ms duration with 10 ms rise- and 10 ms fall time, ISI between 2 and 5.5 s) generated by a PC-stimulator with Creative Labs Soundblaster 16, were presented to both ears in a pseudorandomized order by headphones. Subjects were asked to press a button with their dominant hand in response to target stimuli. Data were collected with a sampling rate of 250 Hz and an analogous bandpass filter (0,16 – 50 Hz). 350 ms prestimulus and 800 ms poststimulus periods were evaluated for 55 target stimuli. For artifact suppression, all trials were automatically excluded from averaging when the voltage exceeded $\pm 100 \mu V$ in any one of the 32 channels at any time point of the averaging period. For each subject, the remaining sweeps were averaged separately for the target stimuli.

13.3 The Linear Inverse Solution

The major advantage of EEG recordings is given in the fact that changes in time may be observed on a millisecond time scale whereas PET or functional MRI studies can only produce results on a time scale of seconds. Furthermore EEG activity is produced directly by the electrical activity with which the brain communicates, rather than the indirect correlates as regional blood flow or glucose metabolites, that are imaged by the alternative functional modalities. The problem to solve is localizing electrical sources within the brain which has attracted the EEG community for some time now.

By tesselating the cortex in N disjoint regions using the Talairach atlas by Talairach and Tournoux (1988) the cortex is divided into $N = 2394$ cubic areas (voxels). For each of these voxels the probability of belonging to a certain anatomic area may be given. Representing the sources in each region by a current dipole oriented normal to the surface with amplitude j_i the EEG inverse problem can be expressed in terms of a linear model. The linear forward model relating the N sources j to the M EEG electrodes can be written as

$$\Phi = Kj + \epsilon$$

where the ith row of the MxN transfer matrix K may be viewed as the projection of the lead field (sensitivity) of the ith electrode. The jth column of K specifies the gain vector for the jth dipole moment. The term ϵ describes noise such as background brain activity or modeling errors. Evidently since $N \gg M$ the Matrix K is not of full rank and thus the simple minimum norm solution $\hat{j} = (K^T K)^{-1} K \Phi$ for $\hat{j}$ is not a suitable solution of the inverse problem. Frequently a Tikhonov regularization (Tikhonov and Arsenin 1977) which introduces a penalty term is applied to regularize the solution. The basic idea is to introduce additional information. Instead of minimizing only the norm $\|\Phi - Kj\|^2$ we introduce additional information using a functional of j and a Lagrange multiplier λ.

$$\|\Phi - Kj\|^2 + \lambda \|Ij\|^2$$

The solution is given by:

$$\hat{j} = \left[K^T K + \lambda I \right]^{-1} K \Phi$$

The parameter λ may be found by trial and error using the trade off curve between a purely data driven solution and a purely functional driven solution. Instead of assuming constant activity by introducing the identity matrix I more elaborate functionals of j are possible. Here we use the solution proposed by Pascal-Marqui et al. (1994) who apply the discrete Laplace operator B to the sources j. The idea here is that neighboring voxels have similar activity, which is quite compatible with neurophysiology. Introducing a diagonal weight matrix W with $w_{ii} = \|K_i\|$ they solve the constrained minimization problem:

$$min\|WBj\|^2$$

under the constraint $\Phi = Kj$ with respect to j. Using Lagrange multipliers the solution is given by:

$$\hat{j} = T\Phi$$

with $T = (WBB^T W)^{-1} K^T (K(WBB^T W)K^T)^{-1} \Phi$

Applying this algorithm to the measured scalp data for a single point in time we obtain the data structure for a given individual as shown in Table 13.1.

13.4 The Mixture Model

13.4.1 Initial Preparation of the Data

Once the inverse solution is determined for each point in time two different time points can be considered in order to quantify the activation due to the auditory stimulus. Let $A_i(l)$ denote the current density before activation

Table 13.1 *Data structure of the inverse solution*

Voxel	x-coord.	y-coord.	z-coord.	$\hat{j}_i(x10^{-3})$
1	-52	-11	-41	1.748
2	-45	-11	-41	1.887
3	-38	-11	-41	1.959
...	...	...	...	...
2394	-17	10	71	3.10

and $B_i(l)$ the current density after the stimulus for the ith subject at the lth voxel. In the present study we chose A to be the baseline before the stimulus and B as the time point 84 milliseconds after the auditory stimulus. That means for each subject we do have a set of independent mean difference images for each subject i, $i = 1, \ldots, n$ at repetition j, $j = 1, \ldots, m$ and at voxels $l = 1, \ldots, N$.

$$\Delta_i(l) = \sum_{j=1}^{m} \frac{1}{m}(B_{ij}(l) - A_{ij}(l)) \; i = 1, \ldots, n, \quad j = 1, \ldots, m \quad l = 1, \ldots, N$$

The mean and variance images of these data are estimated as $\bar{\Delta}(l)$ and $S^2(l)$ respectively:

$$u_l := \bar{\Delta}(l) = \frac{1}{n} \sum_{i=1}^{n} \Delta_i(l)$$

$$\sigma_l^2 := S^2(l) = \frac{1}{n-1} \sum_{i=1}^{n} [\Delta_i(l) - \bar{\Delta}(l)]^2$$

Thus at each voxel l we do have a mean value with corresponding variance averaged over the n individuals.

13.4.2 Formulation of the Mixture Model

The first question mentioned in the introduction, i.e. is there any activation translates into investigation of unobserved heterogeneity. This implies that there are subpopulations of voxels which show different activation after stimulus. A natural choice to investigate unobserved population heterogeneity is given in finite mixture models. Frequently we assume a density $f(j, \mu)$ for the phenomenon of interest. The parameter μ denotes the parameter of the population, whereas u is in the sample space U, a subset of the real line. Frequently this model is too strict and a natural assumption would be the *heterogeneous* case, where we assume that the population of interest consists of several subpopulations denoted by $\mu_1, \mu_2, \ldots, \mu_k$. In contrast to the homogenous case we have the same type of density for each subpopulation but different parameters μ_j in the jth subpopulation. In the sample $u_1, u_2, \ldots, u_N$ it is not observed to which subpopulation the information belongs. Therefore this phenomenon is called *unobserved heterogeneity*. Now let a latent variable Z describe population membership. Then the joint density $f(u, z)$ may be written as:

$$f(u,z) = f(u|z)f(z) = f(u|\mu_z)p_z \tag{13.1}$$

where $f(u|z)$ is the density conditionally on membership in subpopulation z. Thus the unconditional density $f(u)$ is given by the marginal density summing over the latent variable

$$f(u, P) = \sum_{j=1}^{k} f(u|\mu_j)p_j \tag{13.2}$$

In this case p_j is the probability of belonging to the j-th subpopulation having parameter μ_j. As a result the p_j are subject to the constraints $p_j \geq 0$ and $p_1 + \cdots + p_k = 1$. Thus (13.2) denotes a mixture distribution with *mixing kernel* $f(u_i|\mu_j, \sigma_i^2)$ and mixing distribution

$$P \equiv \begin{bmatrix} \mu_1 & \cdots & \mu_k \\ p_1 & \cdots & p_k \end{bmatrix} \tag{13.3}$$

in which weights $p_1, \ldots, p_k$ are given to parameters $\mu_1, \ldots, \mu_k$. Here we define $f(u_i, \mu_j, \sigma_i^2) = \frac{1}{\sqrt{2\pi\sigma_i^2}} \exp(-\frac{1}{2}(u_i - \mu_j)^2/\sigma_i^2)$. Thus in analogy to meta-analysis we consider the individual voxel as a study of over n individuals and use the normal density with fixed variance σ_i^2 considered as known. Estimation of the parameters of the mixing distribution P is predominantly done using maximum-likelihood (ML). Given a sample of:

$$u_i \overset{iid}{\sim} f(u|P), \quad i = 1, \ldots, N$$

we are interested in finding the ML estimates of P which maximize the log likelihood function:

$$l(P) = \log L(P) = \sum_{i=1}^{N} \log \sum_{j=1}^{k} f(u_i, \mu_j, \sigma_i^2)p_j$$

Please note that also the number of components k needs to be estimated as well. Two cases must be distinguished: In the *flexible support size* case P may vary in the set of all probability measures Ω. This guarantees that Ω is a convex set and that $l(P)$ is a concave functional on the set of all discrete probability measures. This is the basis for the strong results of nonparametric mixture distributions. An efficient algorithmic solution is given by the VEM algorithm which looks on a fixed grid of parameters for grid points with positive support without the need to make an initial assumption about the number of components k. Details may be found in Böhning

Table 13.2 *ML estimates of the finite mixture model*

Weights $\hat{p}_j$	Parameter $\hat{\mu}_j$ (x10^{-3})
0.0062	0.380
0.0269	0.667
0.0589	0.952
0.1089	1.282
0.1564	1.639
0.2443	2.084
0.2616	2.615
0.1212	3.350
0.0156	4.652

et al. (1992) and Böhning (1999). The solution of the VEM-algorithm may then be used as starting values for the EM-algorithm by Dempster et al. (1977b) which gives a solution for the fixed support size case.

Applying the mixture algorithm, i.e. the combination of the VEM and the EM-algorithm as outlined before leads to a mixture distribution with ten components. In order to estimate the number of components we apply the nonparametric bootstrap, as described by Schlattmann and Böhning (1997). This leads to a final solution with nine components as shown in Table 13.2. The estimation of the mixture model has been done with the software C.AMAN (Böhning et al. 1992, Böhning 1999) which can handle in its UNIX-version large data sets as the present one with 2394 data points.

The first question of interest in activation experiments deals with heterogeneity of activation. Clearly for these data we find heterogeneity of activation. There are a total of nine subpopulations ranging from a subpopulation mean of 0.38x10^{-3} to 4.652x10^{-3}. Evidently, there are about 14% of the voxels which show a rather strong activation after 84 ms post auditory stimulus.

13.5 Classification of the Inverse Solution

The second objective of neurophysiological activation studies is to identify the activation status of the individual voxel. Classification of the individual voxel is straightforward in the finite mixture model setting. The ith voxel may simply be categorized applying Bayes' theorem as follows:

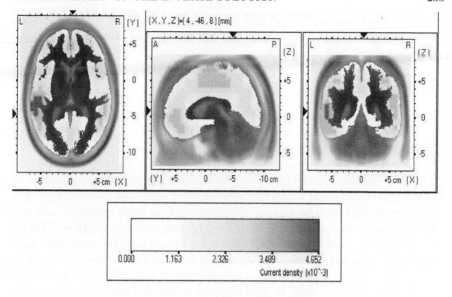

Figure 13.1 *Talairach slice of the acoustic cortex*

$$Pr_{\boldsymbol{P}}(Z_{ij} = 1 \mid U_i = u_i) = \frac{Pr(U_i = u_i \mid Z_{ij} = 1)Pr(Z_{ij} = 1)}{\sum_l Pr(U_i = u_i \mid Z_{il} = 1)Pr(Z_{il} = 1)}$$

$$= \frac{\hat{p}_j f(u_i, \hat{\mu}_j, \sigma_i^2)}{\sum_l \hat{p}_l f(u_i, \hat{\mu}_l, \sigma_i^2)}$$

The individual voxel is now categorized into the category for which the probability of belonging becomes maximal. Figure 13.1 shows a slice out of the Talairach atlas where the classification has been performed in this way. Due to the relatively large number of components and the property of the software LORETA only to provide a continuous scale for the display of the data we focus on the areas with high activity.

In terms of interpretation we find several areas which show strong activation which are consistent with neurophysiological knowledge. The source analysis of the N1-component revealed two main areas of activity namely the superior temporal lobe on both sides, which contain the primary and secondary auditory cortex. This localization is in agreement with intracerebral recordings in humans (Knight et al. 1988) as well as neuro-imaging studies (Tzourio et al. 1997) employing auditory stimulation. The greater left than right activity observed in the present data is in line with a previous magnetencephalographic study (Elberling et al. 1982) and may be due to the known anatomical asymmetry of the superior temporal cortex with a larger area on the left side (Galaburda and Sanides 1980) The presented method is a useful tool linking surface measured potentials to their generating cerebral structures.

13.6 Discussion

The statistical evaluation of activation experiments is quite important in neuro-imaging. For functional imaging techniques such as PET, functional MRI and electrophysiological methods based on the inverse solution there is the need to identify heterogeneity in space after an activation in time. The finite mixture model presented here allows a model-based classification of the individual voxel which requires only few assumptions due to its semi-parametric nature. The mixing distribution P is nonparametric and only the mixing kernel $f(x, \mu, \sigma^2)$ requires a parametric assumption. This approach to the classification of the individual voxel may be performed unsupervised, i.e. using the mixture algorithm and the nonparametric bootstrap the analysis may be performed in an automated way. Likewise the stability of the classification may be investigated using the parametric bootstrap (McLachlan and Basford 1988). This approach can also be applied to the data of a single individual. This allows in contrast to other methods classification of individual PET or MRI images. In this case the variance of the individual subpopulations needs to be estimated. In general the likelihood is unbounded in this case, thus either a common variance is fixed or estimates based on local maxima of the EM-algorithm can be applied. However, the properties of these estimates still need to be investigated thoroughly.

One drawback of the method is given by the fact, that it assumes independence of the observations. Although heterogeneity may mimic autocorrelation and vice versa, as is well known from the area of disease mapping, no particular correlation structure can be modeled. Work is in progress to extend the methodology to hidden Markov random fields which allows the introduction of more sophisticated correlation structures.

CHAPTER 14

Spatio-Temporal Cluster Modelling of Small Area Health Data

A.B. Clark and A.B. Lawson

14.1 Introduction

The spatial analysis of small area health data is a topic which has developed greatly in the last decade. In this development, there has been considerable interest in the analysis of clustering of disease. This has come about in response to a need within public health to be able to analyse disease maps with a view to establishing whether clustering exists, particularly where a possible environmental hazard may be related to the adverse disease outcome. Much of the methodology developed in this area has focussed on hypothesis testing and little has focussed on statistical modelling of clustering. Recent reviews of these developments can be found in Lawson and Kulldorff (1999), and Lawson (2001), Chapter 6. There are many advantages to modelling of clusters, not least of which is the flexibility to introduce covariates and to be able to include a range of descriptive parameters within the fitting process. While relatively little development of spatial models has been witnessed, the analysis of clustering in spatio-temporal small area health data has seen even less development.

In Chapter 1 of this volume, an introduction of the concepts of cluster modelling are presented.

14.2 Basic Cluster Modelling approaches

In this section we examine some basic approaches to clustering. A basic working definition of clustering is given by Knox (1989) as: '*a geographically bounded group of occurrences of sufficient size and concentration to be unlikely to have occurred by chance*'. In this definition, it is implied that closeness of cases of disease is a criteria which is submitted to a statistical test of some kind. The definition is based on case event data, i.e. where the locations of cases of disease are known and are to be analysed. Here we will consider two basic data forms: case events and counts of disease cases within arbitrary administrative units (such as census tracts or zip codes). In Knox's definition there is no restriction on the form of the 'cluster'. In the situation where counts are to be considered then an equivalent definition could be: '*a geographically bounded group of regions with disease counts of sufficient size and concentration to be unlikely to have occurred by chance*'. It is also implicit in the definition that spatial variation in the population 'at risk' from the disease of interest is taken into consideration within the statistical testing procedure.

Here, as well as distinguishing between case event and count data, we also introduce the distinction between methods which assess the overall clustering tendency of the disease map (general or *non-specific* clustering) and those methods which aim to assess where clusters occur on maps (*specific* clustering). Our focus in cluster modelling is on *specific* clustering and methods for modelling the locations of clusters.

14.2.1 Case Event Data Models

For case event data, often the spatial distribution of the population 'at risk' is represented by the distribution of a *control* disease. In this situation, many cluster testing procedures use distance measures between cases and neighbouring cases, or between cases and control disease events as their basic tool for describing the clustering behaviour of the cases. In the situation where counts of disease are examined, it is the level of count and levels within neighbouring regions that determine whether a group of regions is regarded as a cluster.

Inter-Event versus Hidden Process Modelling

Initially we will consider a disease map consisting of spatial coordinates of the location of cases of disease. Often such case event data can be considered to form a point process, and further it is also assumed that a conditional first order intensity can be specified for the events of the process. This intensity which is usually assumed to be continuous over space, specifies the local aggregation of cases on the map, and can be a function of a variety of factors. For example, spatial covariates may affect the distribution of cases (e.g. environmental pollution gradients). Alternatively, the intensity could be defined at an arbitrary location of s conditionally as a function of a hidden process of cluster centres, or of the locations cases around the arbitrary location. We define the intensity at s as:

$$\lambda(s|\{\mathbf{c}_j\}, \{\mathbf{x}_i\}, \theta), \tag{14.1}$$

where $\{\mathbf{x}_i\}$ is the realisation of n cases, $\{\mathbf{c}_j\}$ is a set of m putative cluster 'centres' and θ is a p−valued parameter vector. We could also include dependence on covariates here but this is not considered in the general specification at this point. We envisage that clustering can be modelled by the appropriate specification of this intensity. In descriptive clustering studies often the degree of local aggregation is assessed via the estimation of the equivalence of a spatial covariance function (for point events this would be a $K(t)$ function (Diggle 1983)). Here, we attempt to consider clustering via the intensity function, the nature of the clustering often being controlled by prior distributions for important clustering parameters. The specification in equation (14.1) can be simplified by considering two general models for clustering: models where dependence is defined by the locations of cases of disease (inter-event models) and models where dependence is with an un-observed process of 'centres' (hidden process models). Inter-event models can be specified by:

$$\lambda(s|\{x_i\}, \theta), \qquad (14.2)$$

whereas hidden process models can be specified by:

$$\lambda(s|\{c_j\}, \theta). \qquad (14.3)$$

For inter-event models the intensity at any case event must be specified conditionally on the other events : $\lambda(x_i|\{x\}_{-i}, \theta)$, where $\{x\}_{-i}$ denotes the case event set with the i th event removed. Notice that the inter-event models defined here are close in form to nonparametric approaches to clustering where, for example, the local density of events is deemed to be a smooth function of the locations of other events (Kelsall and Diggle 1998). A variant of these models could specify dependence on control disease events, but we assume here that variation in the population background is included elsewhere within the parameterisation of the $\lambda(.)$ function.

An example of an inter-event model which could be envisaged is:

$$\lambda(s|\{x_i\}, \theta) = \rho.g(s).[\eta + \sum_{\{x_i\}} h(s - x_i; \theta)/g(x_i)]. \qquad (14.4)$$

Here, $g(s)$ is a function representing the background 'at risk' population, η is a constant (usually 1.0) which allows the null hypothesis of cases distributed from the background variation alone, $h(.)$ is a prespecified cluster distribution function which describes the relationship between locations and events. Clearly in this specification the localised intensity is a function of the spatial aggregation of the realisation of cases. Modifications of the summation component could be envisaged where nearest neighbours or neighbours within a distance range could be included only in the sum. In fact the distance range (δ_x) for the neighbourhood sum could be regarded as a clustering parameter, thus:

$$\lambda(s|\{x_i\}, \theta) = \rho.g(s).[\eta + \sum_{\{x_i \in \delta_x\}} h(s - x_i; \theta)/g(x_i)]. \qquad (14.5)$$

Another possible variant, is to hypothesise that the intensity in (14.4) is valid and the θ parameters consist, at least in part, of a set of cluster spread parameters $\{\kappa_i\}$ $i = 1, ..., n$, one for each observation. Here the spread parameters, describing localised aggregation of cases, would be unknown and have to be estimated.

Estimation for these inter-event models is complicated by the fact that for any realisation the likelihood is not readily available due to the conditioning on the other events. It is possible to formulate a Bayesian hierarchical model for these intensities, and this can help to identify model components but the conditioning in the likelihood doesn't allow simple progress. One possibility is to assume a pseudo-likelihood for the data, given that small area disease clusters are likely to be weakly expressed and that ignoring the correlation in the pseudo-likelihood should not be serious for such weak correlation.

The alternative form of cluster model, that of the hidden process model, is typically specified thus:

$$\lambda(s|\{c_j\}, \theta) = \rho.\mathbf{g}(s).[\eta + \sum_{j=1}^{m} h(s - c_j; \theta)].$$

where ρ is a scaling parameter representing the relative frequency of cases to controls. Here, the $\{c_j\}$ form a hidden process, usually a point process, so that they represent unobserved locations. Usually both the $\{c_j\}$ and m are unknown and must be estimated.

The idea of a hidden process is not limited to a set of point locations. Instead, some other form of object process could be considered. For example, if clusters were considered to be of linear form, then a line segment process could be included. Further still, a mixture of hidden processes could be conceived (e.g. point and line clusters representing diffuse and road or riverside clusters, or a non-specific random effect component and a cluster process).

Other forms of hidden process could be considered. For example, it is possible to consider that the non-spatial marginal distribution of risk consists of a mixture of risk levels and a map of risk can be classified into well-defined risk classes or partitions. For count data, this has been proposed by Schlattmann and Böhning (1993) and Knorr-Held and Rasser (2000). It has also been used to provide a nonparametric trend model for counts (Denison and Holmes (2001), see also Chapter 7 in this volume). In the above models it is often natural or necessary to introduce prior distributions for the components of models (e.g. the number and location of cluster centres; the number of and level of risk in partitions). Hence, it is often necessary to design a Bayesian hierarchical modelling approach to cluster analysis. Posterior sampling via Markov chain Monte Carlo (MCMC) algorithms is often employed and for mixture problems with unknown numbers of components (cluster locations or risk levels), reversible jump (RJ) or birth-death (BD) MCMC usually has to be employed (Knorr-Held and Rasser 2000, Stephens 2000, Cressie and Lawson 2000).

14.2.2 Small Area Count Data Models

For small area count data a variety of clustering models can be envisaged. The inter-event models of equations (14.4-14.5) can be modified straightforwardly to the count case. For example, for $i = 1, ..., n$ regions, one version of model (14.4) could be defined for counts as:

$$E(y_i) = \rho e_i [\alpha + \sum_{j=1}^{n} \frac{y_j}{e_j} h(\boldsymbol{x}_i - \boldsymbol{x}_j; \kappa_j)],$$

where $\boldsymbol{x}_i$ is the centroid of the i th region, and y_i, e_i the observed and expected count in the i th region respectively, and clustering is controlled by κ_j parameters assigned locally.

Partition models have also been proposed as specific cluster models, although they appear to be better suited to nonparametric trend estimation, as they do not parameterise cluster locations themselves. A further possibility is to conceive of groupings of regions controlled by a prior distribution which controls the cluster form. This has been advocated by Gangnon and Clayton (2000) and Chapter 8, in this volume.

14.2.3 Spatio-Temporal Extensions to Cluster Models

In principle, it is straightforward to extend many of the cluster models in Section 14.2 to spatio-temporal settings. For example, if the approach utilising first order intensities is used, then we can define a spatio-temporal first order intensity as

$$\lambda(\boldsymbol{s}, t | \{\boldsymbol{x_i}, u_i\}, \{\boldsymbol{\Phi}\}, \theta),$$

where $\{\boldsymbol{x_i}, u_i\}$ are the set of case events indexed by location ($\boldsymbol{x_i}$) and time of occurrence/diagnosis (u_i), and $\{\boldsymbol{\Phi}\}$ is a set of cluster terms which may be defined for spatial, or temporal or spatio-temporal cluster effects. Many interesting issues arise in this context concerning appropriate parameterisation of these terms, as there could potentially be a large number of interactions between space and time. Finally an interesting feature of models with temporal components is the fact that conditioning on previous observed events is possible. For example, the inter-event models used could be defined as either

$$\lambda(\boldsymbol{s}, t | \{\boldsymbol{x_i}, u_i\}, \{\boldsymbol{\Phi}\}, \theta),$$

or

$$\lambda(\boldsymbol{s}, t | \{\boldsymbol{x_i}, u_i : u_i < t\}, \{\boldsymbol{\Phi}\}, \theta),$$

or a combination of these dependencies could be assumed, depending on whether the model exploits this conditioning or not. In the next section a hidden process space-time model is defined and applied to a birth abnormality case study.

14.3 A Spatio-Temporal Hidden Process Model

In this section, we propose a model that can be used to study the spatio-temporal distribution of small area health data (disease incidence or prevalence) either at individual case level (i.e. date and location of the cases known) or at some aggregate level where the count of disease is only available within a space-time unit. The model proposed here, has components which represent localised increases in excess risk (i.e. clustering).

Current methodology for the assessment of spatio-temporal clustering dates back to the work of Knox (1964) who essentially reduced the problem to a 2 way contingency table consisting of the numbers of cases and controls that are cross-classified as being close in time and being close in space for a predefined distance. Mantel (1967) generalised Knox's test by defining temporal closeness and spatial closeness for a number of distances and then summing over the distances. Recent work by Diggle and coworkers (Diggle et al. 1995) who use Ripley's $K(t)$ function is similar in spirit to Knox's test. More recently Kulldorff and Nagarwalla (1995) have developed a spatio-temporal scan test. However, all these tests have disadvantages. For example, they cannot allow for covariates, Knox's test requires an arbitrary spatial and temporal lag to be defined, and they are subject to bias from the choice of study region. In addition, with the exception of the scan test, they are designed for the assessment of overall spatio-temporal clustering and not the detection of the locations of clusters in space-time.

The methods developed here are based on spatio-temporal mixture models, and are illustrated using a data set of post-coded birth abnormalities in Tayside, Scotland, between January 1991 and December 1995.

14.4 Model Development

In this case study, we consider the situation where the residential location of cases and the time of birth is known and an appropriate point process model is considered. These define the coordinates of the locations in space-time as $\{\boldsymbol{x}_i, u_i\}$. We consider a realisation of n such case events $\{\boldsymbol{x}_i, u_i\} i = 1, ..., n$ within spatial window A and temporal window T. We assume that the data can be modelled using a modulated heterogeneous Poisson process with first order intensity $\lambda(\mathbf{s}, t)$. With this intensity we need to take account of the underlying variation in population. For example, if we were dealing with Sudden Infant Death Syndrome then we would need to allow for the differing number of live births throughout the study region. This modulation, or 'at risk' *background*, is assumed to act multiplicatively on the intensity. Thus we assume an intensity of the form

$$\lambda(\boldsymbol{s}, t) = \mathbf{g}(\boldsymbol{s}, t).r(\boldsymbol{s}, t; \mathbf{z}), \qquad (14.6)$$

where $\mathbf{g}(.,.)$ is the background of the disease, $r(.,.;\mathbf{z})$ is the relative risk function at $\{s,t\}$ which can depend on measured covariates $\mathbf{z}$, which may (or may not) be spatially defined, or on underlying cluster centre locations (unspecified for brevity). The particular parameterisation of $r(.)$ depends upon the application of interest. For example, for focused clustering $\mathbf{z}$ could include the distance and angle from a putative health hazard (Lawson 1993b, Diggle 1990) but might also include some temporal component. Notice that $\mathbf{g}(.,.)$ is a nuisance function of which we must take account, in order to make reasonable inferences about $\lambda(s,t)$, but which is not the primary focus of the analysis.

For clustering we consider the intensity to be increased around notional cluster centres. Our aim is to estimate these putative centres both in their locations and number. We consider these centres to be the primary underlying source of heterogeneity in the excess risk. This approach is distinct from general random effect modelling of spatio-temporal risk variation (see e.g. Zia et al. (1997), Waller et al. (1997b), and also Chapters 11 and 12 in this volume) which assumes that heterogeneity is a random disturbance in the level of the intensity. Often this disturbance is modelled by a log Gaussian prior distribution within a hierarchical Bayesian formulation. That approach cannot directly model locations of clusters and also makes global assumptions about the nature of heterogeneity which may not always be tenable, e.g. in sparse disease incidence studies.

We can envisage three different types of increased risk in our cluster model formulation:

1. The disease may increase for a period of time and then return to its previous level throughout the whole spatial study region. We shall call this a *temporal cluster*.

2. The disease may be elevated in a region of space throughout the whole study period. We shall call this a *spatial cluster*.

3. The disease may be elevated in a region of space and for a period of time. We shall call this a *spatio-temporal cluster*.

Each of these types of cluster is defined by its *persistence* properties, that is, a temporal cluster must persist over the full study region, a spatial cluster must persist over the full time period and a spatio-temporal cluster is non-persistent. This has implications for the choice of both study region and period. If the study region is too large we may not detect a temporal cluster and if the period is too long we may not detect a spatial cluster. It is for these reasons that the choice of study period and region is still of importance in a modelling framework just as with a hypothesis testing framework.

If we further assume that these three cluster types act independently and additively on the first order intensity of the process we may write

$$\lambda(s,t) = \mathbf{g}(s,t).\exp(\beta^T \mathbf{z})m\{\underline{\mathbf{c}}_1, \underline{\mathbf{c}}_2, \underline{\mathbf{c}}_3, s,t\},\qquad(14.7)$$

where

$$m\{\underline{\mathbf{c}}_1, \underline{\mathbf{c}}_2, \underline{\mathbf{c}}_3, s,t\} = \{1 + \alpha_1 \sum_{i=1}^{n^s} K(s - \mathbf{c}_{1i}) + \alpha_2 \sum_{i=1}^{n^t} K(t - \mathbf{c}_{2i})$$

$$+ \alpha_3 \sum_{i=1}^{n^{st}} K((s,t) - \mathbf{c}_{3i})\},$$

where $\{\mathbf{c}_{1i}\}, i = 1, ..., n^s$ are the spatial cluster centres, $\{\mathbf{c}_{2i}\}, i = 1, ..., n^t$ are the temporal cluster centres, $\{\mathbf{c}_{3i}\}, i = 1, ..., n^{st}$ are the spatio-temporal cluster centres. The 'weights' $\{\alpha_j\}, j = 1, 2, 3$ represent the increase in relative risk invoked by the cluster centres. In addition, $\beta^T \mathbf{z}$ is a linear predictor, a function of additional covariates, which can be included in the model formulation. These covariates could be spatial or temporal trend, or, for example area level deprivation indices. The cluster distribution function $K(.)$ describes the density of cases around the centre. This can take a variety of forms. For example, we may assume that it is more likely for cases to occur near the cluster centre. The constant term (one) allows the intensity to reduce to the 'at risk' background level as we get further away from the cluster centres (in both space and time). Notice also that as we get further away in time, both the temporal and spatio-temporal components decay to zero and we are left with pure spatial clustering.

The formulation above has the advantage of providing a flexible modelling strategy for different types of clustering. For example, if it were thought the data only exhibited temporal clustering we could set $\alpha_1 = \alpha_3 = 0$. Further, because we examine posterior samples of the cluster centres and number of centres, within a Bayesian sampling algorithm, we can examine a variety of posterior configurations of number and location of centres, and overlays of these realisations can provide evidence for different spatial and temporal scales of clustering. Further extensions of this approach could include cluster distribution functions which have spatially-dependent variances which allow variable sizes of clusters, or the inclusion of conventional random effects (in addition to cluster terms). The extensions have been described for the spatial case by Lawson (1995) and Lawson (2000). However, in this paper, we confine our attention to the simpler case where clusters have a constant variance parameter (though the variance is sampled as a parameter in the posterior sampling algorithm), and no additional random effects are admitted.

The properties of this model and its associated inference, in space, are discussed at length in Lawson and Clark (1999b). It should also be noted that the application of this type of modelling is not restricted to epidemiology but can also be made to problems in geosciences (Cressie and Lawson 2000) and the approach may also have application in other fields such as astrostatistics.

14.4.1 Estimation of $\mathbf{g}(s,t)$

The estimation of the background intensity, $\mathbf{g}(s,t)$, must be considered since it is a nuisance function of which we need take account. It is natural to consider some nonparametric estimate of $\mathbf{g}(s,t)$, as we do not wish to make inferences on $\mathbf{g}(s,t)$. This estimate is often based on the spatial distribution of some measure of the 'at risk' background. In early work on focused clustering, Diggle (1990) and Lawson and Williams (1994) suggested conditioning on such an estimated background, i.e. a plug-in estimate was used. However, subsequent inference did not take account of the underlying variability in this estimation. In response to this problem, Diggle and Rowlingson (1993) proposed a model based on the location of cases and the distribution of the locations of a control disease, and suggested a regression model which conditioned out the $\mathbf{g}(s,t)$ from the analysis.

Lawson and Clark (1999a) proposed a Bayesian method which allows the exploration of the full joint posterior distribution of the smoothing parameter in the nonparametric estimator of $\mathbf{g}(s,t)$. Lawson and Williams (1994) compared, in the spatial case, the use of a control disease with the use of a population-based estimate of risk and concluded that the results obtained differed.

It should be noted that all of these methods assume that the background smoothing is constant over the whole study region. However, if a study region is split into highly populated areas (e.g. cities) and sparsely populated areas (e.g. rural areas) then it might be more appropriate to model the smoothing adaptively e.g. as a nearest neighbour estimator. Similar problems exist in multivariate density estimation near the edge of the parameter space, Simonoff (1993) Chapter 4.

If the spatial distribution of a control disease is available, it is matched to the case disease based on the age-sex 'at risk' distribution of the case disease, but unaffected by the phenomenon under investigation. Then it is possible to estimate the space-time background function $\mathbf{g}(s,t)$ nonparametrically as in the spatial case. In that case we could employ: a nonparametric density estimate of the space-time variation in the control disease, such as

$$\hat{\mathbf{g}}(s,t) = \sum_j w_1 \left(\frac{s - \mathbf{v}_j}{h_1} \right) w_2 \left(\frac{t - \tau_j}{h_2} \right), \qquad (14.8)$$

where w_1 and w_2 are kernel functions, h_1 and h_2 are smoothing parameters and $\{\mathbf{v}_j, \tau_j\}$ is the space-time coordinates of a realisation of a control disease. Note that in this formulation there is a different smoothing constant in time and space but that there is equal smoothing in both spatial directions. It is well known that such multivariate kernel density estimation suffers from the 'curse of dimensionality'. For that reason, other simpler alternatives might be considered necessary. If the disease is rare, and there is little information about any space-time interaction, one possibility is to consider a separable estimator of the form:

$$\hat{\mathbf{g}}(s,t) = g_{01}(s).g_{02}(t) = \sum_j w_1 \left(\frac{s - \mathbf{v}_j}{h_1} \right) . \sum_j w_2 \left(\frac{t - \tau_j}{h_2} \right). \qquad (14.9)$$

An estimator as specified as (14.8) or (14.9) can be substituted into the model (14.7) to fully specify the intensity. A likelihood, conditional on observing n events, can be derived for this model and can be expressed as

$$L = \prod_{i=1}^n \hat{\mathbf{g}}(\boldsymbol{x_i}, u_i).r(\boldsymbol{x_i}, u_i; \mathbf{z}). \left(\int_A \int_T \hat{\mathbf{g}}(\mathbf{p}, q)r(\mathbf{p}, q; \mathbf{z})dqd\mathbf{p} \right)^{-n}. \qquad (14.10)$$

The Prior Distribution for Cluster Centres

The cluster centres themselves form a point process, however this process is not observed. In this model formulation there are three cluster components, each with unknown number and locations of centres. This can be represented as a mixture problem with an unknown number of components and component locations. Without further ancillary information, assumptions must be made to allow the estimation of such parameters. It is conventional that spatial components such as cluster centres have spatial prior distributions and the number of centres must also have some prior distribution. It is possible to assume a spatially uniform prior distribution for locations and a Poisson(δ) distribution for the number, say. Alternatively, to prevent reconstructions with multiple response (in our sampling algorithms), some inhibition in the prior distribution of centres could be assumed, see van Lieshout and Baddeley (1995) for further discussion of multiple response problems. Indeed cluster centre estimation will not be efficient or 'convergent' without (slight) inhibition of centres. A Strauss prior distribution (see for example Chapter 4 in this volume) provides a joint prior distribution for both the locations and number of centres. The Strauss prior distribution, for the spatial case, is defined for a realisation of m centres, thus:

$$f(\underline{\mathbf{c}}) \propto b^m \gamma^{R(\underline{\mathbf{c}})}, \qquad (14.11)$$

where, b is a rate parameter, γ an inhibition parameter, R an inhibition distance and $R(\underline{\mathbf{c}})$ is the number of $R - close$ pairs in the realisation of centres. There are three parameters in this definition. If such prior distributions are employed for all cluster types, independently, there will be nine parameters. Both R and γ, which control the degree of inhibition, are highly correlated, and are usually fixed at values representing weak inhibition. The rate parameter, b, can also be fixed. The parameters values assumed for these parameters are detailed in Section 14.6.3. Other choices of the prior distribution can be envisaged and are discussed, at length, in Chapter 4 in this volume.

Choice of Cluster Distribution Function

The cluster distribution function describe how cases are spread around a cluster centre. A variety of possibilities exist for this function. It is common to use a radial distribution function with a Gaussian form, although a uniform distribution within a disc is also possible. The variance parameter of the cluster distribution function determines the spatial scale or spread of the clusters. Here we assume these to be constant for each cluster of the same type. The practical effects of this are that cluster centres located in the built-up areas will dominate (since the clusters will be less dispersed than in rural areas) and hence we may not pick up centres located in rural areas.

It should be noted that the choice of cluster distribution function is, in all probability, of secondary importance to the estimation of the variance parameter. The radial symmetric Gaussian distribution is used throughout and is given by

$$K(\mathbf{s} - \mathbf{c}) = \frac{1}{2\pi\kappa} \cdot \exp\{-\|\mathbf{s} - \mathbf{c}\|^2 / 2\kappa\}, \qquad (14.12)$$

where κ is the cluster variance, see Chapter 4 of this volume.

Other Prior Distributions and the Posterior Distribution

With the likelihood given by equation (14.10) and with the background intensity incorporated the prior distributions that need to be specified are for the cluster centres (location and number), and the weights. For the weight vector it is convenient to adopt reasonably uninformative and independent exponential prior distributions:

$$g(\alpha_1, \alpha_2, \alpha_3) = \exp(-\alpha_1 - \alpha_2 - \alpha_3), \qquad (14.13)$$

i.e. each weight is exponentially distributed with mean 1. Note that the weights are not constrained to sum to one. When covariates are included in the model, then the regression parameters associated with these are assumed to have uninformative uniform prior distributions on suitable ranges. For the location and number of cluster centres we adopt the Strauss inhibition prior distributions:

$$
\begin{aligned}
\pi(\underline{\mathbf{c_1}}) &\propto b_1^{n^s} \gamma^{R(\underline{\mathbf{c}}_1)}, \\
\pi(\underline{\mathbf{c_2}}) &\propto b_2^{n^t} \gamma^{R(\underline{\mathbf{c}}_2)}, \\
\pi(\underline{\mathbf{c_3}}) &\propto b_3^{n^{st}} \gamma^{R(\underline{\mathbf{c}}_3)},
\end{aligned}
\qquad (14.14)
$$

where b is a rate parameter, γ is an interaction parameter and $R(\underline{c})$ is the number of R-close pairs in $\underline{c}$. As mentioned previously one of the main uses of such priors is to prevent multiple centres having the same location. It is possible to assume a joint Strauss inhibition prior for the spatial cluster centres and the spatial part of the spatio-temporal cluster centres, similarly we can specify the joint prior for the temporal cluster centres and the temporal part of the spatio-temporal cluster centres. This formulation would prevent spatial and spatio-temporal cluster centers existing at the same location in space and similarly would prevent temporal and spatio-temporal cluster centers existing at the same point in time. However, it was found that this did not unduly effect the results in Section 14.6 and is hence not applied.

The resulting joint posterior distribution for the model can be specified as proportional to:

$$\frac{\prod_{i=1}^{n} \left[\hat{\mathbf{g}}(\boldsymbol{x_i}, u_i) \exp(\beta^T \mathbf{z}_i)\{1 + M_1(\boldsymbol{x_i}) + M_2(u_i) + M_3((\boldsymbol{x_i}, u_i))\}\right]}{\{\int_A \int_T \hat{\mathbf{g}}(\mathbf{p}, q) \exp(\beta^T \mathbf{z})(1 + M_1(\mathbf{p}) + M_2(q) + M_3((\mathbf{p}, q)) dq d\mathbf{p}\}^{-n}}$$
$$\times \ \exp(-\alpha_1 - \alpha_2 - \alpha_3) \ \times \ b_1^{n^s} \gamma^{R(\underline{c}_1)} \ \times \ b_2^{n^t} \gamma^{R(\underline{c}_2)} \ \times \ b_3^{n^{st}} \gamma^{R(\underline{c}_3)}.$$

Where

$$M_1(\boldsymbol{s}) \ = \ \alpha_1 \sum_{j=1}^{n^s} K(\boldsymbol{s} - \mathbf{c}_{1j}),$$

$$M_2(t) \ = \ \alpha_2 \sum_{j=1}^{n^t} K(t - \mathbf{c}_{2j}),$$

$$M_3((\boldsymbol{s}, t)) = \alpha_3 \sum_{j=1}^{n^{st}} K((\boldsymbol{s}, t) - \mathbf{c}_{3j}).$$

Notice that the complexity of the problem increases drastically if we allow any of the Strauss parameters to vary, since the normalising constant is a function of the Strauss parameters. Possible solutions to this include using various approximations to the Strauss prior distributions. This is, however, not pursued here. In Section 14.5, we discuss the algorithm used to sample the centres and related parameters. The formulation used here is specified in Section 14.6.

14.4.2 Region Counts

In the data example examined here, the sparseness of the incidence of cases and lack of orderliness supports the aggregation of cases into counts within small areas. Here we have aggregated to postcode sectors (5 digits) and calendar months. Because of this aggregation, we have considered modelling the situation where count data aggregation of the basic point process model is employed.

Count data are derived from aggregating case-event data to arbitrary regions. This type of data is routinely collected and normally used for large scale studies (e.g. cancer atlases). A large number of models exist for dealing with this type of data (see e.g. Lawson (2001)). Most of these models make the assumption that we observe a count within a spatio-temporal cell, say $\{y_{it}\}$ $i = 1, ..., m,$ $t = 1, ..., T,$ and a set of expected counts, say $\{e_{it}\}$ $i = 1, ..., m,$ $t = 1, ..., T.$ The distribution of the counts is usually assumed to be:

$$y_{it}|\{\theta_{it}\} \backsim \text{Poisson}(e_{it}.\theta_{it}), \qquad (14.15)$$

and modelling concentrates on the relative risks $\{\theta_{it}\}$. These models, although widely used, have been the subject of recent criticism (for example Lawson and Cressie (2000)). In this section we extend the case-event model of the previous section to deal with count events aggregated into space-time cells.

Integrated Intensity

If the cases form a heterogeneous modulated Poisson process with intensity given by equation (14.6), then the number of cases in arbitrary (non overlapping) regions $\{A_i\}$ and time intervals $\{\delta_t\}$ are independent Poisson random variables with parameter

$$m_{it} = \Lambda(A_i, \delta_t|\underline{c}_1, \underline{c}_2, \underline{c}_3) = \int_{A_i} \int_{\delta_t} \lambda(\mathbf{s}, t)dtd\mathbf{s}. \qquad (14.16)$$

This approach has a number of advantages over model (14.15).

1. It does not involve the definition of a neighbourhood structure, which is often assumed for conventional count data models.
2. It takes account of both the size and shape of the study region.
3. In model (14.15), a decoupling approximation is made, i.e. whereby it is assumed that everyone lives at the centroid of each region, see Lawson and Cressie (2000) for more details.

The integration is easily done if we have a control disease for which we have the exact spatial and temporal location. However this will rarely occur in practice and we will normally have just the count in the region. This means that we must approximate the integral as

$$m_{it} \simeq e_{it} \exp(\beta^T \mathbf{z}). \int_{A_i} \int_{\delta_t} \{1 + \alpha_1 M_1(\mathbf{s}) + \alpha_2 M_2(t) + \alpha_3 M_3((\mathbf{s},t))\}dtd\mathbf{s}$$

Conditional on the total count (say N) the likelihood is then of multinomial form.

14.5 The Posterior Sampling Algorithm

The use of iterative sampling algorithms is now widespread in complex statistical modelling. The most popular technique, Markov chain Monte Carlo (MCMC), is a method which allows one to iteratively draw samples from the posterior distribution. Given the sample values it is then possible to draw inferences about the joint or marginal distributions of each parameter.

The extension of MCMC algorithms to the situation where the parameter space varies is not straightforward. However, Geyer and Møller (1994), Green (1995) and Richardson and Green (1997) have suggested possible methodology. Models fitted here can be viewed as variable parameter mixture models. For such mixture problems, drawing inferences from these chains is more difficult since you may visit components a relatively small number of times.

The model presented here consists of a number of parameters which can be updated using standard MCMC techniques (Carlin and Louis 1996). However, for both case event or count data applications, the cluster centres need to be updated using variable parameter space MCMC techniques, such as reversible jump MCMC. For our problem, this can be done using a birth-death-shift algorithm as described in Geyer and Møller (1994) and summarized below.

Consider a set of points $\underline{\mathbf{v}} = (\mathbf{v}_1, .., \mathbf{v}_n)$ then an algorithm using iterative updates via births and deaths of points can be specified. Updates for this can be made set by one of the following steps

1. *BIRTH: (or addition)* The addition of a randomly selected new point. The location of the new point (say $\mathbf{u}$) generated from a density $b(\mathbf{u}|\underline{\mathbf{v}})$ (that usually depends on the location of the other points).

2. *DEATH: (or deletion)* The removal of an existing point (say $\mathbf{v}_i$), the point is chosen from a distribution $d(\mathbf{v}_i|\underline{\mathbf{v}})$.

3. *SHIFT: (or random displacement)* The random movement of one point (say $\mathbf{v}_i$), this is chosen from the current set with equal probability and is moved to $\mathbf{u}$ with density $s(\mathbf{v}_i, \mathbf{u})$.

The birth (B) -death (D) -shift (S) steps are incorporated in a Metropolis-Hastings algorithm where each type of transition (B,D,S) has a probability of being chosen. Here, q is defined as the birth probability and p is defined as the probability of a shift transition. The transitions (B,D,S) are defined as :$(1-p)qb(\mathbf{u}|\underline{\mathbf{v}}); (1-p)(1-q)d(\mathbf{v}|\underline{\mathbf{v}}); ps(\mathbf{u}, \mathbf{v})$ respectively.

The Metropolis-Hastings acceptance ratios for these moves are given by

$$BIRTH : LR \times PR \times \frac{1 - q(\underline{\mathbf{v}} \cup \mathbf{u})}{q(\underline{\mathbf{v}})} \cdot \frac{d(\mathbf{u}|\underline{\mathbf{v}} \cup \mathbf{u})}{b(\mathbf{u}|\underline{\mathbf{v}})},$$

$$DEATH : LR \times PR \times \frac{q(\underline{\mathbf{v}} \setminus \mathbf{v}_i)}{1 - q(\underline{\mathbf{v}})} \cdot \frac{b(\mathbf{v}_i|\underline{\mathbf{v}} \setminus \mathbf{v}_i)}{d(\mathbf{v}_i|\underline{\mathbf{v}})},$$

$$SHIFT : LR \text{ x } PR \text{ x } \frac{s(\mathbf{u}, \mathbf{v}_i)}{s(\mathbf{v}_i, \mathbf{u})},$$

where LR is the likelihood ratio, and PR is the prior ratio. This algorithm, and other steps used for other parameters are detailed in the Appendix. Extensions to using exact sampling techniques, such as those described in Chapter 5 in the volume, are certainly possible but are not pursued here.

14.5.1 Goodness-of-Fit Measures for the Model

As a model choice criterion the *Bayesian Information Criterion (BIC)* is widely used in Bayesian and hierarchical models as it asymptotically approximates a Bayes factor. In a model with log-likelihood $l(\theta)$ the BIC value is estimated from the output of an MCMC algorithm by

$$2\hat{l}(\theta^*) - p \ln n,$$

where p is the number of parameters, n is the number of data points and

$$\hat{l}(\theta^*) = \frac{1}{G} \sum_{i=1}^{G} l(\theta_i^*),$$

the averaged log-likelihood over G posterior samples of θ. In our model we do not have a fixed number of parameters. Here we replace p by $\hat{p}$, the average number of parameters in the G posterior samples of θ. We can thus consider this setup as calculating a BIC value for every posterior sample and averaging the value over the sample.

Notice that, recently, the *Deviance Information Criterion* (DIC) has been proposed by Spiegelhalter et al. (1999). This is defined as

$$DIC = 2E_{\theta|x}\{D\} - D\{E_{\theta|x}(\theta)\},$$

where $D(.)$ is the deviance of the model and x is the observed data. This uses the average of the posterior samples of θ to produce an expected value of θ. However, if this parameter is a spatial (or temporal) location over varying dimensional space then the average may not be a useful summary measure of the posterior distribution.

14.6 Data Example: Scottish Birth Abnormalities

14.6.1 Introduction

The data set that we consider is the location of birth abnormalities in Tayside (Scotland) between January 1991 and December 1995. The birth abnormalities examined consists of all abnormalities recorded at birth on hospital births records (Scottish Morbidity Record scheme: SMR2 database). While abnormalities vary in both type and aetiology, the variation in space and time of total abnormalities may give indications as to possible differentials in health service provision or in environmental exposure risk. Potential material exposure to harmful environmental agents, lifestyle variation and diagnostic practice are possible determinants. Following aggregation, the data form a set of counts in months (60) and in postcode sectors (92). In addition to the counts of abnormalities, also available is the count of live births for the same units. The live birth count is used to provide an estimate of the variation in the population background, which we need to take account of when assessing the excess variation in the abnormality count. Deprivation indices for the post code sectors were also available (Carstairs 1981). These may yield evidence for lifestyle variation or deprivation links.

14.6.2 Exploratory Analysis

A time series plot of the standardised mortality ratio (SMR)* is given below (Figure 14.1) and shows only slight evidence of a possible increase in time, but also suggests two possible clusters around 20 and 50 months.

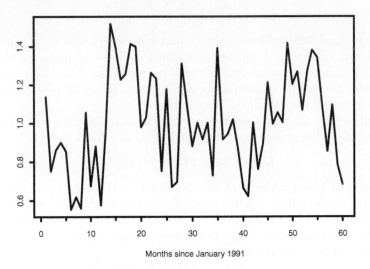

Months since January 1991

Figure 14.1 *A time series plot of the Standardised Mortality Ratio, defined as $SMR_t = m_t * \sum n_t / \sum m_t$ where m_t is the number of live births in month t and n_t is the number of birth abnormalities in month t.*

This suggests that a cluster model in the temporal marginal distribution could be appropriate.

* The SMR mortality ratio for the ith region is here defined as the ratio of the observed count to the expected count, where the expected count is $(m_{it}/m..) * n...$. The dot stands for summation over the relevant subscript.

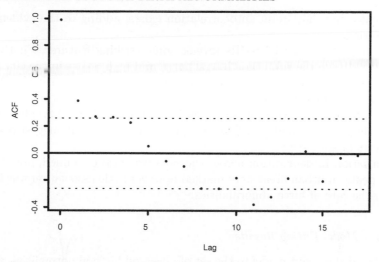

Figure 14.2 *The autocorrelation function of temporal for each month from 1/1/91 until 31/12/95 SMR_t with 5% significance level.*

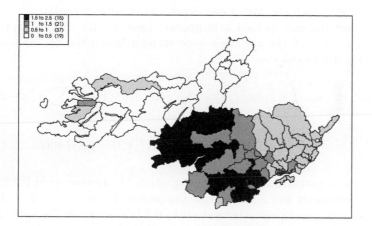

Figure 14.3 *A plot of the spatial SMR in Tayside 1991-1995 defined as $SMR_i = m_i * \sum n_i / \sum m_i$.*

The temporal structure of the number of live births and the number of birth abnormalities is different in a number of respects. Firstly, the number of births is declining with time whereas the number of abnormalities is not. Secondly, the number of live births shows evidence of seasonality whereas the number of abnormalities does not. Also the temporal autocorrelation function of the SMR (Figure 14.2) reveals that although no seasonality is present some short term autocorrelation exists, adding to the evidence of clustering.

A map of the spatial SMRs reveals some striking features. Firstly, the small SMRs in the rural (north-west) area and high SMRs in the city areas (Dundee and Perth). Secondly, elevated SMRs tend to cluster in the middle of the map (Figure 14.3).

The SMRs for each year are given in Figures 14.4 to 14.8. These show that the SMR are highly variable in space-time and might suggest possible clusters in space-time.

Variation in deprivation across the area was also examined, not shown, and peaks in urban areas were marked however little association was found with the rate of birth abnormalities.

14.6.3 Model Fitting Results

We fitted the count model to the set of observed birth abnormalities, where the space-time units are postcode sector and calendar month. The control used was the number of live births in the same space-time unit. Following the exploratory analysis, we included the Carstairs deprivation index (dep), and x coordinate and the y coordinate as possible covariates for spatial trend, but did not include a temporal trend as this did not appear to be important. All the variables were standardised prior to analysis. The resulting full model was:

$$
m_{it} \simeq e_{it} \cdot \int_{A_i} \int_{\delta_t} \exp\{\beta_1 x + \beta_2 y + \beta_3.dep\}\{1 + \alpha_1 \sum_{k=1}^{n^s} K(\boldsymbol{x} - \mathbf{c}_{1k})
$$

$$
+ \alpha_2 \sum_{k=1}^{n^t} K(t - \mathbf{c}_{2k}) + \alpha_3 \sum_{k=1}^{n^{st}} K((\boldsymbol{x},t) - \mathbf{c}_{3k})\}dtd\boldsymbol{x}. \qquad (14.17)
$$

The Strauss parameters (b_i and γ in equation 14.14) were set at 0.2 for each radius parameter and 3 for the rate parameter (i.e. $b_1 = b_2 = b_3 = 3$). The inhibition distance parameter was set to 0.1 (measured on the unit square). These settings were motivated by prior beliefs concerning the range of potential inhibition and exploratory analysis of the form of cluster variation possible. These settings lead to relatively strong inhibition of centres, which we would favour here as we wish to ensure well separated cluster estimates and we believe a priori that a small number of centres exist in the pattern. The weights were given exponential prior distributions with mean 1 as these were assumed a priori to take values less than 1, and the β parameters were given uniform prior distributions on large ranges, as we have no strong prior preference for the values of these parameters.

The Metropolis-Hastings steps were run until convergence, which was usually achieved within the first 20,000 iterations, and then for a further

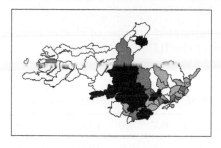

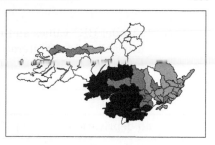

Figure 14.4 *A plot of the SMR in Tayside for 1991, with SMR and legend as defined in Figure 14.3.*

Figure 14.5 *A plot of the SMR in Tayside for 1992, with the SMR and legend as defined in Figure 14.3.*

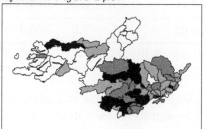

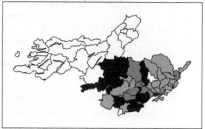

Figure 14.6 *A plot of the SMR in Tayside for 1993, with SMR and legend as defined in Figure 14.3.*

Figure 14.7 *A plot of the SMR in Tayside for 1994, with the SMR and legend as defined in Figure 14.3.*

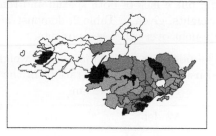

Figure 14.8 *A plot of the SMR in Tayside for 1995, with the SMR and legend as defined in Figure 14.3.*

period of time. Multiple starting points were examined. Convergence monitoring was carried out using Geweke's criterion and also parameter summaries. Details of the algorithm are given in the appendix. The posterior sample estimates are based on the last 500 iterations, after convergence. A variety of subset models were examined (fitted separately). In Table 1 the results of fitting subsets with one component removed are presented.

Table 1: Model BIC values and the estimated number of cluster centres
using the last 500 iterations of the RJMCMC algorithm

Model	$2\hat{l}(\theta^*)$	n^s	n^t	n^{st}	BIC
full	-27740.36	6.8	4.0	9.0	-55712.7
no deprivation	-27723.70	6.7	5.0	14.0	-55718.8
no x-coordinate	-27712.79	8.6	2.1	10.0	-55656.8
no y-coordinate	-27785.42	7.8	2.1	6.4	-55766.6
no spatial clustering	-28007.71	-	5.0	6.4	-56163.6
no temporal clustering	-27759.02	6.7	-	9.0	-55700.9
no spatio-temporal clustering	-27781.62	6.0	2.1	-	-55684.9

This table shows some interesting features. Notice how the number of
spatial and temporal clusters change only slightly in each model, whereas
the number of spatio-temporal clusters changes dramatically.

Our results indicate a lack of any relationship between the level of deprivation of an area and the number of birth abnormalities within that area.
A strong gradient from the west coast to the east, strong evidence spatio-
temporal clustering, but a lack of both spatial and temporal clustering.

The estimated weights, given in Table 2, demonstrate the importance of
the temporal component over other components.

Table 2 : The posterior expectation of the weight parameters
estimated using the last 500 iterations of the RJMCMC
Algorithm

Model	α_1	α_2	α_3
full	0.57	2.02	0.17
no deprivation	0.75	1.92	0.12
no x-coordinate	1.05	1.32	0.37
no y-coordinate	1.88	0.80	0.38
no spatial clustering	0.00	2.66	0.03
no temporal clustering	0.22	0.00	0.05
no spatio-temporal clustering	0.05	0.16	0.00

This table demonstrates that the temporal clustering pattern is relatively

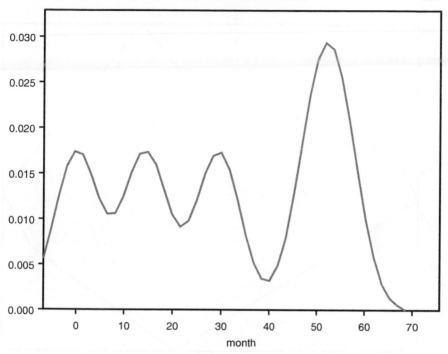

Figure 14.9 *A kernel smoothed estimate of the locations of cluster centres in time, based on the last 500 iterations.*

unchanged in any of the models, although in the no-spatial-clustering model we see that its importance increases.

The location of the cluster centres in time is given in Figure 9. This shows that the most likely location of a cluster is after 50 months, and the next two most likely locations near the start of the temporal window. This may however be an artifact of the intensity estimation method. The location of the cluster centres in space is given in Figure 10. This shows a cluster in the north-west of Tayside and other possible centres in rural locations.

The results compare well with the exploratory analysis. The temporal peaks in the SMR (Figure 14.3) are not present in the results although neither of these two peaks has an SMR which differs significantly from 1.0. The regression parameters also show the urban/rural effect and little relationship between deprivation and birth abnormalities. It should be noted however that without the deprivation index as a covariate we have a greater number of spatio-temporal clusters.

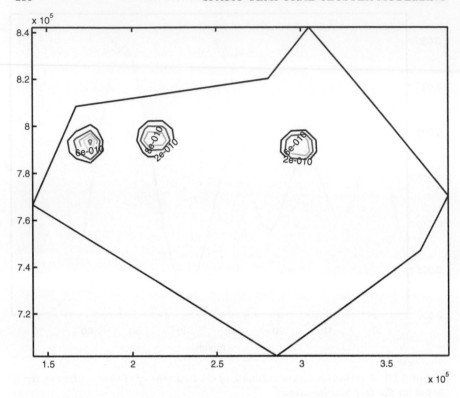

Figure 14.10 *A kernel smoothed estimate of the locations of cluster centres in space, based on the last 500 iterations. The bounding polygon is an approximation to the external bounding polygon of the study region.*

14.7 Discussion

We have introduced a new model for the analysis of small area health data at both the case-event level or aggregated count level. This model allows a general framework for clustering in the spatio-temporal domain. With only slight modifications it can include known cluster centres, e.g. waste incinerators, it can also allow for the examination of different types of clustering avoiding any multiple testing problems. We have also discussed how one can assess the goodness-of-fit of a clustering model.

While one can never replace a carefully designed (and executed) case-control study with any observational data, the development of realistic models for clustering can suggest the need for further study. The inclusion of covariates into such models is of prime importance, indeed in our formulation the cluster terms represent unmeasured covariates. However, it is often difficult to obtain covariates which can represent unobserved heterogeneities and resort has to be made to crude measures such as the Kafadar-Tukey urbanisation index or Cairstairs deprivation index. The numerical values of covariates is usually hard to obtain, usually one has covariates measured at some aggregate level which may not correspond to the subdivision by postcodes. For such cases useful models for covariate measurement error would be needed.

Within the modelling framework presented here further refinements are possible, for example one could remove the need for a constant cluster variance and allow the clusters to vary in size in the relevant domain of interest. We have not considered edge effects here, but these can be taken into account by weighting the likelihood or using an external guard area.

The model here is computationally intensive, but given recent advances in computer technology we do not feel that this is a major drawback. Also the extension of this work into the development of a spatio-temporal surveillance alarm system would be of great public health importance and is a long term research aim.

Our results compare favourably with the results of previous studies on birth abnormalities, for example Dolk et al. (1998). They found little evidence of an association with deprivation and a rural-urban gradient, albeit of a different form. Some limitations of our study are, firstly, we are dealing with total abnormalities without differentiation into different types. This is the subject of future work. We have also not looked at the differences between health authority areas (which could lead to differential diagnosis/treatment outcomes). Clearly in a fuller study such differences could also be included in the analysis.

It should also be noted that the models proposed here are specifically designed for the assessment of the location and number of clusters in space, time and space-time whereas other studies in space-time disease incidence examine random effect models for the variation (e.g. Xia and Carlin 1998). While these models can account for extra-variation and correlation in the relative risk, they do not provide a mechanism, and are not designed, for the estimation of cluster locations. Those models are appropriate for the general modelling of levels of variation in disease incidence.

In summary, we believe that the methodology proposed here can provide a relevant basis for cluster studies and can provide enhanced capability to detect cluster locations and number.

Appendix: Metropolis-Hastings BDS algorithm

The algorithm introduced in Section 14.5 depended upon the choice of a birth density, a death distribution and a shift density. The mixing of the algorithm is dependent upon both the choice of these functions and the proportion of time we choose to update using a birth, death or shift.

While all these may depend upon the number of points and their location, for simplicity, we have chosen the following:

1. We propose a birth, death or shift with equal probability, i.e. $p = 0.33, q = 0.5$ and

2. If we propose a birth we propose the new location at random throughout the study area unit.

3. If we propose a death we choose the location from the current set with equal probability.

4. If we propose a shift we choose a point from the current set with equal probability and shift it to a new location chosen at random on the study area unit.

With this choice the ratios are, for a generic set of cluster centres $\{c_i\}, i = 1, ..., n^c$, given by

$$
\begin{aligned}
\text{Birth} &: LR \times PR \times 2 \times \frac{1}{n^c}, \\
\text{Death} &: LR \times PR \times \frac{1}{2} \times n^c, \\
\text{Shift} &: LR \times PR.
\end{aligned}
$$

Notice that the birth ratio decreases as n^c increases and the death ratio increases as n^c increases.

Our overall algorithm, at each step of the chain, is:

1. update $\alpha_1, \alpha_2, \alpha_3$ using a Gaussian jumping kernel centered around the current value and a fixed width,

2. update $\beta_1, \beta_2, \beta_3$ using a Gaussian jumping kernel centered around the current value and small variance,

3. update $\underline{c}_1$ using the birth death shift algorithm defined above,

4. update $\underline{c}_2$ using the birth death shift algorithm defined above,

5. update $\underline{c}_3$ using the birth death shift algorithm defined above.

Acknowledgements

The authors wish to acknowledge the support of an EU Biomed2 concerted action grant, number BMH4-CT96-0633, and the Information and Statistics Division of the National Health Service (Scotland) for the provision of the data.

References

Adler, R. (1981). *The Geometry of Random Fields*, Wiley, New York.

Ahrens, C., Altman, N., Casella, G., Eaton, M., Hwang, J. T. G., Staudenmayer, J. and Stefanscu, C. (1999). Leukemia clusters and TCE waste sites in upstate New York: How adding covariates changes the story, *Technical report*, School of Operations Research, Cornell University.

Allard, D. and Fraley, C. (1997). Non-parametric maximum likelihood estimation of features in spatial point process using Voronoi tessellation, *J. Amer. Statist. Assoc.* **92**: 1485–1493.

Arak, T., Clifford, P. and Surgailis, D. (1993). Point-based polygonal models for random graphs, *Adv. Appl. Prob.* **25**: 348–372.

Aykroyd, R. G. (1995). Partition models in the analysis of autoradiographic images, *Appl. Statist.* **44**: 441–454.

Azzalini, A. and Capitanio, A. (1999). Statistical applications of the multivariate skew normal distribution, *J. Roy. Statist. Soc. B* **61**: 579–602.

Azzalini, A. and Dalla Valle, A. (1996). The multivariate skew-normal distribution, *Biometrika* **83**: 715–726.

Baddeley, A. J. (1999). Spatial sampling and censoring, *in* O. Barndorff-Nielsen, W. S. Kendall and M. N. M. van Lieshout (eds), *Stochastic geometry, likelihood, and computation*, Boca Raton: Chapman & Hall.

Baddeley, A. J. and Møller, J. (1989). Nearest-neighbour Markov point processes and random sets, *International Statistical Review* **57**: 89–121.

Baddeley, A. J. and van Lieshout, M. N. M. (1993). Stochastic geometry models in high-level vision, *in* K. V. Mardia and G. K. Kanji (eds), *Statistics and Images, Advances in Applied Statistics, a supplement to the Journal of Applied Statistics*, Vol. 20, Carfax Publishing, Abingdon, chapter 11, pp. 231–256.

Baddeley, A. J. and van Lieshout, M. N. M. (1995). Area-interaction point processes, *Annals of the Institute of Statistical Mathematics* **46**: 601–619.

Baddeley, A. J., Fisher, N. I. and Davies, S. J. (1993). Statistical modelling and prediction of fault processes, *in* P. J. Hatherly, J. Shepherd, B. J. Evans and N. I. Fisher (eds), *Integration of methods for the prediction of faulting*, Sydney: Australian Coal Industry Research, chapter 6, pp. 143–176.

Baddeley, A. J., Møller, J. and Waagepetersen, R. P. (2000). Non- and semi-parametric estimation of interaction in inhomogeneous point patterns, *Statistica Neerlandica* **54**: 329–350.

Baddeley, A. J., van Lieshout, M. N. M. and Møller, J. (1996). Markov properties of cluster processes, *Advances in Applied Probability* **28**: 346–355.

Banfield, J. D. and Raftery, A. E. (1993). Model-based Gaussian and non-Gaussian clustering, *Biometrics* **49**: 803–821.

Barry, D. and Hartigan, J. A. (1993). A Bayesian analysis for changepoint problems, *J. Amer. Statist. Assoc.* **88**: 309–319.

Bartlett, M. S. (1975). *The Statistical Analysis of Spatial Pattern*, Chapman and Hall, London.

Bendrath, R. (1974). Veralgemeinerung eines Satzes von R.K. Milne, *Mathematische Nachrichten* **59**: 221–228.

Benes, V., Bodlak, K., Møller, J. and Waagepetersen, R. P. (2001). Bayesian analysis of log Gaussian Cox process models for disease mapping, In preparation.

Bensmail, H., Celeux, G., Raftery, A. E. and Robert, C. P. (1997). Inference in model-based cluster analysis, *Statist. Comp.* **7**: 1–10.

Bernardinelli, L., Clayton, D. G., Pascutto, C., Montomoli, C., Ghislandi, M. and Songini, M. (1995). Bayesian analysis of space-time variation in disease risk, *Statistics in Medicine* **14**: 2433–2443.

Bernardo, J. M. and Smith, A. F. M. (1994). *Bayesian Theory*, Chichester: Wiley.

Besag, J. E. (1994). Discussion on the paper by Grenander and Miller, *Journal of the Royal Statistical Society* **B 56**: 591–592.

Besag, J. E., York, J. C. and Mollié, A. (1991). Bayesian image restoration, with two applications in spatial statistics (with discussion), *Ann. Inst. Statist. Math.* **43**: 1–59.

Best, N. G. and Wakefield, J. C. (1999). Accounting for inaccuracies in population counts and case registration in cancer mapping studies, *Journal of the Royal Statistical Society* **162**: 363–382.

Best, N. G., Arnold, R., Thomas, A., Waller, L. A. and Conlon, E. (1998). Bayesian models for spatially correlated disease and exposure data, *in* J. Bernardo, J. Berger, A. Dawid and A. Smith (eds), *Bayesian Statistics 6*.

Best, N. G., Ickstadt, K. and Wolpert, R. L. (2000). Spatial Poisson regression for health and exposure data measured at disparate resolutions, *Journal of the American Statistical Association* **95**: 1076–1088.

Binder, D. A. (1978). Bayesian cluster analysis, *Biometrika* **65**: 31–38.

Bithell, J. (1990). An application of density estimation to geographical epidemiology, *Statistics in Medicine* **9**: 691–701.

Blackwell, P. G. (1998). Bayesian inference for a random tessellation process, *Technical report*, Probability and Statistics Report 485/98, University of Sheffield.

Böhning, D. (1999). *C.A.MAN-Computer Assisted Analysis of Mixtures and Applications*, Chapman and Hall.

Böhning, D., Dietz, E. and Schlattmann, P. (1998). Recent developments in computer assisted mixture analysis., *Biometrics* **54**: 283–303.

Böhning, D., Schlattmann, P. and Lindsay, B. G. (1992). C.A.MAN- computer assisted analysis of mixtures: Statistical algorithms., *Biometrics* **48**: 283–303.

Bookstein, F. L. (1989). Principal warps: thin-plate splines and the decomposition of deformations, *IEEE Transactions on Pattern Analysis and Machine Intelligence* **11**: 567–585.

Bookstein, F. L. (1991). *Morphometric Tools for Landmark Analysis: Geometry and Biology*, Cambridge University Press, Cambridge.

Box, G. E. P. and Cox, D. R. (1969). An analysis of transformations, *J. Roy. Statist. Soc. B* **26**: 211–246.

Breslow, N. and Clayton, D. G. (1993). Approximate inference in generalised linear mixed models, *Journal of the American Statistical Association* **88**: 9–25.

Brix, A. (1999). Generalized gamma measures and shot-noise Cox processes, *Advances in Applied Probability* **31**: 929–953.

Brix, A. and Chadoeuf, J. (2000). Spatio-temporal modeling of weeds and shot-noise G Cox processes. Submitted.

Brix, A. and Diggle, P. J. (2001). Spatio-temporal prediction for log-Gaussian Cox processes, *Journal of the Royal Statistical Society*. To appear.

Brix, A. and Møller, J. (2001). Space-time multitype log Gaussian Cox processes with a view to modelling weed data, *Scandinavian Journal of Statistics* **28**: 471–488.

Brown, P. E., Kaaresen, K. F., Roberts, G. O. and Tonellato, S. (2000). Blur-generated non-separable space-time models, *Journal of the Royal Statistical Society, Series B* **62**: 847–860.

Byers, J. A. (1992). Dirichlet tessellation of bark beetle spatial attack points, *J. Anim. Ecol.* **61**: 759–768.

Byers, S. D. and Besag, J. E. (2000). Inference on a collapsed margin in disease mapping, *Statistics in Medicine* **19**: 2243–2249.

Byers, S. D. and Raftery, A. E. (1998). Nearest neighbor clutter removal for estimating features in spatial point processes, *J. Amer. Statist. Assoc.* **93**: 577–584.

Cai, Y. and Kendall, W. S. (1999). Perfect implementation of simulation for conditioned Boolean model via correlated Poisson random, *Technical report*, Department of Statistics, Warwick University.

Carlin, B. P. and Louis, T. A. (1996). *Bayes and Empirical Bayes Methods for Data Analysis*, Chapman and Hall, London.

Carlin, B. P., Gelfand, A. E. and Smith, A. F. M. (1992). Hierarchical Bayesian analysis of changepoint problems, *Appl. Statist.* **41**: 389–405.

Carstairs, V. (1981). Small area analysis and health service research, *Community Medicine* **3**: 131–139.

Carter, D. S. and Prenter, P. M. (1972). Exponential spaces and counting processes, *Z. Wahr. verw. Geb.* **21**: 1–19.

Casella, G., Mengersen, K. L. and Robert, C. P. (1999). Perfect slice samplers for mixtures of distributions, *Technical report*, Department of Statistics, Glasgow University.

Chatfield, C. and Collins, A. J. (1980). *Introduction to multivariate analysis*, London: Chapman & Hall.

Chaudhuri, P. and Marron, J. S. (1999). SiZer for exploration of structure in curves, *J. Amer. Statist. Assoc.* **94**: 807–823.

Chaudhuri, P. and Marron, J. S. (2000). Scale space view of curve estimation, *Ann. Statist.* **28**: 408–428.

Cheng, M. Y. and Hall, P. (1997). Calibrating the excess mass and dip tests of modality, *J. Roy. Statist. Soc. B* **60**: 579–589.

Chessa, A. G. (1995). *Conditional simulation of spatial stochastic models for reservoir heterogeneity*, PhD thesis, Technical University of Delft.

Chib, S. and Greenberg, E. (1995). Hierarchical analysis of sur models with extensions to correlated serial errors and time-varying parameter models, *J. Econometrics* **68**: 339–360.

Chilès, J. P. (1989). Modèlisation géostatistique de réseaux de fractures, *in* M. Armstrong (ed.), *Geostatistics*, Dordrecht: Kluwer, pp. 57–76.

Christensen, O. F. and Waagepetersen, R. P. (2001). Bayesian prediction of spatial count data using generalised linear mixed models, Submitted.

Christensen, O. F., Møller, J. and Waagepetersen, R. P. (2000). Geometric ergodicity of Metropolis-Hastings algorithms for conditional simulation in generalised linear mixed models, *Technical Report R-00-2010*, Department of Mathematical Sciences, Aalborg University. *Methodology and Computation in Applied Probability*. To appear.

Chu, Y., Guerrero, M., Gruendl, R. A., Williams, R. M. and Kaler, J. B. (2001). Chandra Reveals the X-Ray Glint in the Cat's Eye, *The Astrophysical Journal (Letters)* **553**: L69–L72.

Clayton, D. G. and Bernardinelli, L. (1992). Bayesian methods for mapping disease risk, *in* P.Elliott, J. Cuzick, D. English and R. Stern (eds), *Geographical and Environmental Epidemiology: Methods for Small-Area Studies*, Oxford University Press.

Clayton, D. G. and Kaldor, J. (1987). Empirical Bayes estimates of age-standardized relative risks for use in disease mapping, *Biometrics* **43**: 671–681.

Clyde, M. (1999). Bayesian model averaging and model search stratergies (with discussion), *in* J. M. Bernardo, J. O. Berger, A. P. Dawid and A. F. Smith (eds), *Bayesian statistics 6*, Oxford: Clarendon Press.

Coles, P. and Jones, B. (1991). A lognormal model for the cosmological mass distribution, *Monthly Notices of the Royal Astronomical Society* **248**: 1–13.

Cox, D. R. (1955). Some statistical models related with series of events, *J. Roy. Statist. Soc. B* **17**: 129–164.

Cox, D. R. and Isham, V. (1980). *Point Processes*, Chapman and Hall, London.

Cressie, N. A. C. (1993). *Statistics for Spatial Data*, 2nd edn, London: Chapman & Hall.

Cressie, N. A. C. and Huang, H.-C. (1999). Classes of nonseparable, spatio-temporal stationary covariance functions, *Journal of the American Statistical Association* **94**: 1330–1340.

Cressie, N. A. C. and Lawson, A. B. (2000). Hierarchical probability models and bayesian analysis of mine locations, *Advances in Applied Probability* **32**(2): 315–330.

Cressie, N. A. C. and Mugglin, A. (2000). Spatio-temporal hierarchical modelling of an infectious disease from (simulated) count data, *in* J. Bethlehem and P. van der Heijden (eds), *Compstat 2000*, Physica verlag, Heidelberg.

Daley, D. J. and Vere-Jones, D. (1988). *An Introduction to the Theory of Point Processes*, Springer-Verlag, New York.

Dasgupta, A. and Raftery, A. E. (1998). Detecting features in spatial point processes with clutter via model-based clustering, *J. Amer. Statist. Assoc.* **93**: 294–302.

Davis, J. E. (2001). Event Pileup in Charged Coupled Devices, *The Astrophysical Journal* p. to appear.

De Bonet, J. S. (1997). Multiresolution sampling procedure for analysis and synthesis of texture images, *Computer Graphics Proceedings*, pp. 361–368.

De Oliveira, V., Kedem, B. and Short, D. S. (1997). Bayesian prediction of transformed Gaussian random fields, *J. Amer. Statist. Assoc.* **92**: 1422–1433.

Deibolt, J. and Robert, C. P. (1994). Estimation of finite mixture distributions through Bayesian sampling, *J. Roy. Statist. Soc. B* **56**: 363–375.

Dempster, A. P., Laird, N. M, and Rubin, D. B. (1977a). Maximum likelihood from incomplete data via the EM algorithm (with discussion), *Journal of the Royal Statistical Society, Series B, Methodological* **39**: 1–37.

Dempster, A. P., Schatzoff, M. and Wermuth, N. (1977b). A simulation study of alternatives to ordinary least squares, *J. Amer. Statist. Assoc.* **72**: 77–106.

Denison, D. G. T., Adams, N. M., Holmes, C. C. and Hand, D. J. (2002a). Bayesian partition modelling, *Comp. Statist. Data Anal.* (to appear).

Denison, D. G. T. and Holmes, C. C. (2001). Bayesian partitioning for estimating disease risk, *Biometrics* **57**: 143–149.

Denison, D. G. T., Holmes, C. C., Mallick, B. K. and Smith, A. F. M. (2002b). *Bayesian Methods for Nonlinear Classification and Regression*, Chichester: John Wiley. (to appear).

Dey, D. K., Ghosh, S. K. and Mallick, B. K. (eds) (2000). *Generalised Linear Models: A Bayesian perspective*, New York: Marcel Dekker.

Diggle, P. J. (1978). On parameter estimation for spatial point processes, *J. Roy. Statist. Soc. B* **40**: 178–181.

Diggle, P. J. (1983). *Statistical analysis of spatial point processes*, London: Academic Press.

Diggle, P. J. (1985a). A kernel method for smoothing point process data, *App. Statist.* **34**: 138–147.

Diggle, P. J. (1985b). A kernel method for smoothing point process data, *Applied Statistics* **34**: 138–147.

Diggle, P. J. (1990). A point process modelling approach to raised incidence of a rare phenomenon in the vicinity of a prespecified point, *Jour. Royal Statist. Soc. A* **153**: 349–362.

Diggle, P. J. and Rowlingson, B. (1993). A conditional approach to point process modelling of elevated risk, *Technical Report MA93/83*, Lancaster University.

Diggle, P. J., Chetwynd, A., Haggvist, R. and Morris, S. (1995). Second-order analysis of space-time clustering, *Statistical Methods in Medical Research* **4**: 124–136.

Diggle, P. J., Fiksel, T., Grabarnik, P., Ogata, Y., Stoyan, D. and Tanemura, M. (1994). On parameter estimation for pairwise interaction processes, *Int. Statist. Rev.* **62**: 99–117.

Diggle, P. J., Tawn, J. A. and Moyeed, R. A. (1998). Model-based geostatistics (with discussion), *Appl. Statist.*

Dolk, H., Bushby, A., Armstrong, B. and Walls, P. H. (1998). Geographical variation in anopthalmia and microthalmia in England,1988-1994, *British Medical Journal* **317**: 905–910.

Donoho, D. (1988). One sided inference about functionals of a density, *Ann. Statist.* **16**: 1390–1420.

Dryden, I. and Mardia, K. V. (1997). *The Statistical Analysis of Shape*, Wiley, New York.

Duda, R. O., Hart, P. E. and Stork, D. G. (2001). *Pattern Classification*, second edn, Wiley, New York.

Duffy, P. H., Bartels, P. H. and Burchfield, J. (1981). Significance probability mapping: An aid in the topographic analysis of brain electrical activity, *Electroencephalogr Clin Neurophysiol* **60**: 455–463.

Elberling, C., Bak, C., Kofoed, B., Lebech, J. and Saermark, K. (1982). Auditory magnetic fields from the human cerebral cortex: location and strength of an equivalent current dipole., *Acta Neurol Scand* **65**: 553–569.

Elliott, J. J. and Arbib, R. S. (1953). Origin and status of the house finch in the eastern united states, *Auk* **70**: 31–37.

Elliott, P., Wakefield, J. C., Best, N. G. and Briggs, D. G. (1999). *Spatial Epidemiology: Methods and Applications*, Oxford: University Press.

Elvis, M., Matsuoka, M., Siemiginowska, A., Fiore, F., Mihara, T. and Brinkmann, W. (1994). An ASCA GIS spectrum of S5 0014 + 813 at $z = 3.384$, *Astrophysical Journal* **436**: L55–L58.

Everitt, B. S., Landau, S. and Leese, M. (2001). *Cluster Analysis*, fourth edn, Arnold, London.

Fan, J. Q. and Gijbels, I. (1996). *Local polynomial modelling and its applications*, London: Chapman and Hall.

Fessler, J. A. and Hero, A. O. (1994). Space-alternating generalized expectation-maximization algorithm, *IEEE Transactions on Signal Processing* **42**: 2664–2677.

Fisher, N. I. and Marron, J. S. (2001). Mode testing via the excess mass estimate, *Biometrika.* (to appear).

Fisher, N. I., Mammen, E. and Marron, J. S. (1994). Testing for multimodality, *Comp. Statist. Data Anal.* **18**: 499–512.

Fraley, C. and Raftery, A. E. (1998). How many clusters? which clustering method? answers via model-based cluster analysis, *Computer Journal* **41**: 578–588.

Freeman, P. E., Graziani, C., Lamb, D. Q., Loredo, T. J., Fenimore, E. E., Murakami, T. and Yashida, A. (1999). Statistical analysis of spectral line candidates in gamma-ray burst GRB 870303, *The Astrophysical Journal* **524**: 753–771.

Friston, K. J., Frith, C. D., Liddle, P. F., Dolan, R. J., Lammertsma, A. A. and Frackowiak, R. S. J. (1990). The relationship between global and local changes in PET scans, *Journal of cerebral blood flow and metabolism* **10**: 458–466.

Galaburda, A. and Sanides, F. (1980). Cytoarchitectonic organization of the human auditory cortex, *J Comp Neurol* **190**: 597–610.

Gallinat, J. and Hegerl, U. (1994). Dipole source analysis. linking scalp potentials to their generating neuronal structures., *Pharmacopsychiatry* **27**: 52–53.

Gangnon, R. E. (1998). *Disease Rate Mapping via Cluster Models*, PhD thesis, University of Wisconsin, Madison WI.

Gangnon, R. E. and Clayton, M. K. (2000). Bayesian detection and modeling of spatial disease clustering, *Biometrics* **56**: 922–935.

Gangnon, R. E. and Clayton, M. K. (2001). A weighted average likelihood ratio test for spatial clustering of disease, *Statistics in Medicine* **20**: 2977–2987.

Gelfand, A. E. (1996). Model determination using sampling-based methods, *in* W. R. Gilks, S. Richardson and D. J. Spiegelhalter (eds), *Markov Chain Monte Carlo in Practice*, London: Chapman & Hall, pp. 145–161.

Gelfand, A. E. and Carlin, B. P. (1993). Maximum likelihood estimation for constrained or missing data models, *Can. J. Statist.* **21**: 303–311.

Gelfand, A. E. and Smith, A. F. M. (1990). Sampling-based approaches to calculating marginal densities, *Journal of the American Statistical Association* **85**: 398–409.

Gelman, A. and Rubin, D. B. (1992). Inference from iterative simulations using multiple sequences (with discussion), *Statistical Science* **7**: 457–472.

Gelman, A., Carlin, J. B., Stern, H. S. and Rubin, D. B. (1995). *Bayesian Data Analysis*, Chapman & Hall, London.

Geman, S. and Geman, D. (1984). Stochastic relaxation, Gibbs distributions and the Bayesian restoration of images, *IEEE Trans. Patt. Anal. Mach. Intelligence* **6**: 721–740.

Geyer, C. J. (1999). Likelihood inference for spatial point processes, *in* O. Barndorff-Neilsen, W. S. Kendall and M. N. M. van Lieshout (eds), *Stochastic Geometry: Likelihood and Computation*, CRC Press, New York, chapter 3.

Geyer, C. J. and Møller, J. (1994). Simulation procedures and likelihood inference for spatial point processes, *Scan. J. Statist.* **21**: 84–88.

Ghosh, M., Natarajan, K., Stroud, T. and Carlin, B. P. (1998). Generalized linear models for small-area estimation, *Journal of the American Statistical Association* **93**: 273–282.

Ghosh, M., Natarajan, K., Waller, L. A. and Kim, D. (1999). Hierarchical Bayes GLMs for the analysis of spatial data: An application to disease mapping, *J. Statist. Plan. Inf.* **75**: 305–318.

Ghosh, S., Nowak, Z. and Lee, K. (1997). Tessellation-based computational methods for the characterization and analysis of heterogeneous microstructures, *Composites Sci. Tech.* **57**: 1187–1210.

Gilks, W. R., Richardson, S. and Spiegelhalter, D. J. (1996). *Markov chain Monte Carlo in practice*, London: Chapman and Hall.

Giudici, P., Knorr-Held, L. and Rasser, G. (2000). Modelling categorical covariates in Bayesian disease mapping by partition structures, *Statist. Med.* **19**: 2579–2593.

Glasbey, C. A. (1998). A spatio-temporal model of solar radiation microclimate, *in* K. V. Mardia, R. G. Aykroyd and I. L. Dryden (eds), *Proceedings in Spatial Temporal Modelling and its Applications*, Leeds University Press, pp. 75–78.

Godtliebsen, F., Marron, J. S. and Chaudhuri, P. (2001a). Significance in scale space, *Technical report*, Department of Statistics, University of North Carolina.

Godtliebsen, F., Marron, J. S. and Chaudhuri, P. (2001b). Significance in scale space for bivariate density estimation, *J. Comp. Graph. Statist.* (to appear).

Good, I. J. and Gaskins, R. A. (1980). Density estimation and bump-hunting by the penalized maximum likelihood method exemplified by scattering and meteorite data (with discussion), *J. Amer. Statist. Assoc.* **75**: 42–73.

Goodall, C. R. and Mardia, K. V. (1994). Challenges in multivariate spatial modelling, *Proceedings 17th International Biometric Conference, Hamilton, Ontario*, pp. 8–12.

Grandell, J. (1976). *Doubly Stochastic Poisson Processes*, Springer-Verlag, Berlin.

Green, P. J. (1995). Reversible jump Markov chain Monte Carlo computation and Bayesian model determination, *Biometrika* **82**: 711–732.

Green, P. J. and Murdoch, D. J. (1999). Exact sampling for Bayesian inference:

towards general purpose algorithms, *in* J. M. Bernardo, J. O. Berger, A. P. Dawid and A. F. M. Smith (eds), *Bayesian Statistics 6*, Oxford: Oxford University Press.

Green, P. J. and Richardson, S. (2000). Spatially correlated allocation models for count data, *Technical report*, Department of Statistics, Bristol University.

Green, P. J. and Sibson, R. (1978). Computing Dirichlet tessellations in the plane, *Comp. J.* **21**: 168–173.

Häggström, O., van Lieshout, M. N. M. and Møller, J. (1999). Characterisation and simulation results including exact simulation for some spatial point processes, *Bernoulli* **5**: 641–659.

Haldorsen, H. H. (1983). *Reservoir characterization procedures for numerical simulation*, PhD thesis, University of Texas, Austin.

Hand, D. J. (1981). *Discrimination and Classification*, Chichester: Wiley.

Handcock, M. S. and Stein, M. L. (1994). An approach to statistical spatial-temporal modeling of meteorological fields, *J. Amer. Statist. Assoc.* **89**: 368–390.

Handcock, M. S. and Wallis, J. R. (1993). A Bayesian analysis of kriging, *Technometrics* **35**: 403–410.

Handcock, M. S. and Wallis, J. R. (1994). An approach to statistical spatial-temporal modeling of meteorological fields (with discussion), *Journal of the American Statistical Association* **89**: 368–390.

Hans, C. M. (2001). *Accounting for Absorption Lines in High Energy Spectra via Bayesian Modelling*, PhD thesis, Harvard University, Dept. of Statistics.

Hans, C. M. and van Dyk, D. A. (2002). Accounting for absorption lines in high energy spectra, *Statistical Challenges in Modern Astronomy III* (Editors: E. Feigelson and G. Babu), Springer–Verlag, New York.

Hartigan, J. A. and Hartigan, P. M. (1985). The DIP test of multimodality, *Ann. Statist.* **13**: 70–84.

Hartigan, J. A. and Mohanty, S. (1992). The RUNT test from multimodality, *J. Classification* **9**: 63–70.

Haslett, J. (1989). Space time modelling in meteorology — a review, *Bulletin of the International Statistical Institute* **51**: 229–246.

Hastie, T. J. and Tibshirani, R. J. (1990). *Generalized Additive Models*, London: Chapman & Hall.

Hastings, W. K. (1970). Monte Carlo sampling methods using Markov chains and their applications, *Biometrika* **57**: 97–109.

Heikkinen, J. and Arjas, E. (1998). Non-parametric Bayesian estimation of a spatial Poisson intensity, *Scan. J. Statist.* **25**: 435–450.

Heikkinen, J. and Arjas, E. (1999). Modeling a Poisson forest in variable elevation: A nonparametric Bayesian approach, *Biometrics* **55**: 738–745.

Higdon, D., Swall, J. and Kern, J. (1999). Non-stationary spatial modeling, *in* J. M. Bernardo, J. O. Berger, A. P. Dawid and A. F. M. Smith (eds), *Bayesian Statistics VI*, Oxford: Clarendon Press.

Hjort, N. L. and Omre, H. (1994). Topics in spatial statistics, *Scan. J. Statist.* **21**: 289–357.

Hodder, I. and Orton, C. (1976). *Spatial analysis in archeology*, Cambridge: Cambridge University Press.

Holmes, A. P., Blair, R. C., Watson, D. G. and Ford, I. (1996). Nonparametric analysis of statistic images from functional mapping experiments, *Journal of cerebral blood flow and metabolism* **16**: 7–22.

Huang, F. and Ogata, Y. (2001). Comparison of two methods for calculating the partition functions of various spatial statistical models, *Australian and New Zealand J. Statist.* **43**: 47–65.

Hurn, M. (1998). Confocal flourescence microscopy of leaf cells: an application of bayesian image analysis, *Appl. Statist.* **47**: 361–377.

Hurn, M., Husby, O. and Rue, H. (2001). Image analysis, *in* P. Green, N. Hjort and S. Richardson (eds), *Highly Structured Stochastic Systems*, Oxford University Press, London, chapter 7.

Hyndman, R. J., Bashtannyk, D. M. and Grunwald, G. K. (1996). Estimating and visualizing conditional densities, *J. Comp. Graph. Statist.* **5**: 315–336.

Izenman, A. J. and Sommer, C. (1988). Philatelic mixtures and multimodal densities, *J. Amer. Statist. Assoc.* **83**: 941–953.

Jones, R. H. and Zhang, Y. (1997). Models for continuous stationary space-time processes, *in* T. G. Gregoire, D. R. Brillinger, P. J. Diggle, E. Russek-Cohen, W. G. Warren and R. D. Wolfinger (eds), *Modelling Longitudinal and Spatially Correlated Data*, Springer-Verlag, New York, pp. 289–298.

Journel, A. G. and Huijbregts, C. J. (1978). *Mining geostatistics*, Academic Press.

Kang, H., van Dyk, D. A., Yu, Y., Siemiginowska, A., Connors, A. and Kashyap, V. (2002). New MCMC methods to address pile-up in the Chandra x-ray observatory, *Statistical Challenges in Modern Astronomy III* (Editors: E. Feigelson and G. Babu), Springer–Verlag, New York.

Kaufman, L. and Rousseeuw, P. J. (1990). *Finding groups in data: An introduction to cluster analysis*, New York: Wiley.

Kelly, F. P. and Ripley, B. D. (1976). On Strauss's model for clustering, *Biometrika* **63**: 357–360.

Kelsall, J. and Diggle, P. J. (1995a). Kernel estimation of relative risk, *Bernouilli* **1**: 3–16.

Kelsall, J. and Diggle, P. J. (1995b). Non-parametric estimation of spatial variation in relative risk, *Statistics in Medicine* **14**: 2335–2342.

Kelsall, J. and Diggle, P. J. (1998). Spatial variation in risk of disease: a nonparametric binary regression approach, *Applied Statistics* **47**: 559–573.

Kendall, W. S. (1998). Perfect simulation for the area-interaction point process, *in* L. Accardi and C. C. Heyde (eds), *Probability towards the year 2000*, World Scientific Press.

Kendall, W. S. and Møller, J. (2000). Perfect simulation using dominating processes on ordered spaces, with application to locally stable point processes, *Adv. Appl. Prob* **32**: 844–865.

Kendall, W. S. and Thönnes, E. (1999). Perfect simulation in stochastic geometry, *Pattern Recognition* **32**: 1569–1586.

Kent, J. T. and Mardia, K. V. (1994). The link between Kriging and thin plate splines, *in* F. P. Kelly (ed.), *Probability, Statistics and Optimization: A Tribute to Peter Whittle*, Wiley, Chichester, pp. 325–339.

Kent, J. T., Mardia, K. V., Morris, R. J. and Aykroyd, R. G. (2000). Procrustes growth models for shape, *Proceedings of the First Joint Statistical Meeting New*

  Delhi, India, pp. 236–238.

Kent, J. T., Mardia, K. V., Morris, R. J. and Aykroyd, R. G. (2001). Functional models of growth for landmark data, *in* K. V. Mardia and R. G. Aykroyd (eds), *Proceedings in Functional and Spatial Data Analysis*, Leeds University Press, pp. 109–115.

Kingman, J. F. C. (1993). *Poisson Processes*, Clarendon Press, Oxford.

Knight, R. T., Scabini, D., Woods, D. L. and Clayworth, C. (1988). The effects of lesions of superior temporal gyrus and inferior parietal lobe on temporal and vertex components of the human AEP, *Electroencephalogr Clin Neurophysiol* pp. 499–509.

Knorr-Held, L. (2000). Bayesian modelling of inseparable space-time variation in disease risk, *Statistics in Medicine* **19**: 2555–2567.

Knorr-Held, L. and Besag, J. E. (1998). Modelling risk from a disease in time and space, *Statistics in Medicine.* to appear.

Knorr-Held, L. and Rasser, G. (2000). Bayesian detection of clusters and discontinuities in disease maps, *Biometrics* **56**: 13–21.

Knox, E. G. (1964). The detection of space-time interactions, *Applied Statistics* **13**: 25–29.

Knox, E. G. (1989). Detection of clusters, *in* P. Elliott (ed.), *Methodology of enquiries into disease clustering*, Small Area Health Statistics Unit, London, pp. 17–20.

Kolaczyk, E. D. (1999). Bayesian multi-scale models for poisson processes, *Journal of the American Statistical Association* **94**: 920–933.

Kraft, R. P., Kregenow, J. M., Forman, W. R., Jones, C. and Murray, S. S. (2001). Chandra Observations of the X-Ray Point Source Population in Centaurus A, *The Astrophysical Journal* **560**: 675–688.

Kühr, H., Witzel, A., Pauliny-Toth, I. I. K. and Nauber, U. (1981). A catalogue of extragalactic radio sources having flux densities greater than 1 Jy at 5 GHz, *Astronomy and Astrophysics, Supplement Series* **45**: 367–430.

Kulldorff, M. and Nagarwalla, N. (1995). Spatial disease clusters:detection and inference, *Statistics in Medicine* **14**: 799–810.

Kyriakidis, P. C. and Journel, A. G. (1999). Geostatistical space-time models: A review, *Mathematical Geology* **31**: 651–684.

Lange, K. and Carson, R. (1984). Em reconstruction algorithms for emission and transmission tomography, *Journal of Computer Assisted Tomography* **8**: 306–316.

Langford, I., Leyland, A., Rashbash, J., Goldstein, H., McDonald, A. and Bentham, G. (1999). Multilevel modelling of area-based health data, *in* A. B. Lawson, A. Biggeri, D. Boehning, E. Lesaffre, J.-F. Viel and R. Bertollini (eds), *Disease mapping and Risk Assessment for Public Health Decision Making*, Wiley, chapter 16, pp. 217–227.

Lantuéjoul, C. (1997). Conditional simulation of object-based models, *in* D. Jeulin (ed.), *Advances in theory and applications of random sets*, World Scientific Publishing, pp. 271–288.

Last, G. (1990). Some remarks on conditional distributions for point processes, *Stochastic Processes and their Applications* **34**: 121–135.

Lawson, A. B. (1993a). Discussion contribution, *Journal of the Royal Statistical*

Society **B** **55**: 61–62.

Lawson, A. B. (1993b). On the analysis of mortality events associated with a prespecified fixed point, *J. Roy. Statist. Soc. A* **156**: 285–298.

Lawson, A. B. (1995). Markov chain Monte Carlo methods for putative pollution source problems in environmental epidemiology, *Statistics in Medicine* **14**: 2473–2486.

Lawson, A. B. (1997). Some spatial statistical tools for pattern recognition, *in* A. Stein, F. W. T. P. de Vries and J. Schut (eds), *Quantitative Approaches in Systems Analysis*, Vol. 7, C. T. de Wit Graduate School for Production Ecology, pp. 43–58.

Lawson, A. B. (2000). Cluster modelling of disease incidence via RJMCMC methods: a comparative evaluation, *Statistics in Medicine* **19**: 2361–2376.

Lawson, A. B. (2001). *Statistical Methods in Spatial Epidemiology*, Wiley, New York.

Lawson, A. B. and Clark, A. (1999a). Markov chain Monte Carlo methods for clustering in case event and count data in spatial epidemiology, *in* M. E. Halloran and D. Berry (eds), *Statistics and Epidemiology: Environment and Clinical Trials*, Springer verlag, New York, pp. 193–218.

Lawson, A. B. and Clark, A. (1999b). Markov chain Monte Carlo methods for putative sources of hazard and general clustering, *in* A. B. Lawson, D. Bohning, A. Biggeri, J.-F. Viel and R. Bertollini (eds), *Disease Mapping and Risk Assessment for Public Health*, Wiley WHO, chapter 9.

Lawson, A. B. and Clark, A. (2001). Spatial mixture relative risk models applied to disease mapping, *Statitsics in Medicine*. in press.

Lawson, A. B. and Cressie, N. A. C. (2000). Spatial statistical methods for environmental epidemiology, *in* C. R. Rao and P. K. Sen (eds), *Handbook of Statistics:Bio-Environmental and Public Health Statistics*, Vol. 18, Elsevier, pp. 357–396.

Lawson, A. B. and Kulldorff, M. (1999). A review of Cluster Detection Methods, *in* A. B. Lawson, D. B'ohning, E. Lesaffre, A. Biggeri, J.-F. Viel and R. Bertollini (eds), *Disease Mapping and Risk Assessment for Public Health*, Wiley, chapter 7.

Lawson, A. B. and Williams, F. L. R. (1993). Applications of extraction mapping in environmental epidemiology, *Statistics in Medicine* **12**: 1249–1258.

Lawson, A. B. and Williams, F. L. R. (1994). Armadale: a case study in environmental epidemiology, *Journal of the Royal Statistical Society A* **157**: 285–298.

Lawson, A. B., Böhning, D., Lessafre, E., Biggeri, A., Viel, J.-F. and Bertollini, R. (eds) (1999). *Disease Mapping and Risk Assessment for Public Health*, Wiley.

Lindeberg, T. (1994). *Scale-Space Theory in Computer Vision*, Dordrecht: Kluwer.

Lindley, D. V. and Smith, A. F. M. (1972). Bayes estimates for the linear model (with discussion), *J. Roy. Statist. Soc. B* **34**: 1–41.

Link, W. A. and Sauer, J. R. (1998). Estimating population change from count data: application to the North American breeding bird survey, *Ecological Applications* **8**: 258–268.

Loizeaux, M. and McKeague, I. W. (2001). Perfect sampling for posterior landmark distributions with an application to the detection of disease clusters, *IMS Lecture Notes - Monograph Series: Volume 37*, pp. 321–331.

Loredo, T. J. (1993). Promise of Bayesian inference for astrophysics, *Statistical Challenges in Modern Astronomy* (Editors: E. Feigelson and G. Babu), Springer-Verlag, New York, pp. 275–306.

Lucas, P. (2001). Expert knowledge and its role in learning bayesian networks in medicine: An appraisal, *in* S. Quaglioni, P. Barahona and S. Andreassen (eds), *Artificial Intelligence in Medicine*, Springer Verlag, New York, pp. 156–166.

Lucy, L. B. (1974). An iterative technique for the rectification of observed distributions, *The Astrophysical Journal* **79**: 745–754.

Lund, J. and Thönnes, E. (2000). Perfect simulation for point processes given noisy observations, *Technical report*, Department of Statistics, University of Warwick.

Lund, J., Penttinen, A. and Rudemo, M. (1999). Bayesian analysis of spatial point patterns from noisy observations, *Technical report*, Department of Mathematics and Physics, The Royal Veterinary and Agricultural University.

Madigan, D. and Raftery, A. E. (1994). Model selection and accounting for model uncertainty in graphical models using Occam's window, *Journal of the American Statistical Association* **89**: 1535–1546.

Mantel, N. (1967). The detection of disease clustering and a generalised regression approach, *Cancer Research* **27**: 209–220.

Mardia, K. V. and Goodall, C. R. (1993). Spatial-temporal analysis of multivariate environmental monitoring data, *in* G. P. Patil and C. R. Rao (eds), *Multivariate Environmental Statistics*, Elsevier, pp. 347–386.

Mardia, K. V. and Jupp, P. E. (2000). *Directional Statistics*, Wiley, Chichester.

Mardia, K. V., Aykroyd, R. G. and Dryden, I. L. (eds) (1999). *Spatial Temporal Modelling and Its Applications, Proceedings of the 18th LASR Workshop*, Leeds University Press.

Mardia, K. V., Goodall, C. R., Redfern, E. J. and Alonso, F. J. (1998). The Kriged Kalman filter (with discussion), *TEST* **7**: 217–252.

Mardia, K. V., Kent, J. T. and Bibby, J. M. (1979). *Multivariate analysis*, London: Academic Press.

Mardia, K. V., Kent, J. T., Little, J. and Goodall, C. R. (1996). Kriging and splines with derivative information, *Biometrika* **83**: 207–221.

Marshall, R. J. (1991). A review of methods for the statistical analysis of spatial patterns of disease, *J. Roy. Statist. Soc. A* **154**: 421–441.

Matérn, B. (1960). Spatial Variation, Meddelanden fran Statens Skogforskningsinstitut, Band 49, No. 5.

Matérn, B. (1986). *Spatial Variation*, Lecture Notes in Statistics. Springer-Verlag, Berlin.

Matheron, G. (1975). *Random sets and integral geometry*, New York: Wiley.

McCullagh, P. and Nelder, J. A. (1989). *Generalized Linear Models (Second edition)*, Chapman & Hall, London.

McLachlan, G. J. and Basford, K. E. (1988). *Mixture models: Inference and Applications to Clustering.*, Statistics: Textbooks and Monographs, 84. New York etc.: Marcel Dekker, Inc. xi, 253 p. .

Meng, X.-L. and van Dyk, D. A. (1997). The EM algorithm – an old folk song sung to a fast new tune (with discussion), *Journal of the Royal Statistical Society, Series B, Methodological* **59**: 511–567.

Metropolis, N., Rosenbluth, A. W., Rosenbluth, M. N., Teller, A. H. and Teller, E. (1953). Equations of state calculations by fast computing machines, *J. Chem. Phys.* **21**: 1087–91.

Miller, D. M. (1984). Reducing transformation bias in curve fitting, *Amer. Statist.* **38**: 124–126.

Milne, R. K. (1970). Identificability for random translations of Poisson processes, *Zeitschrift für Wahrscheinlichkeitstheorie und verwandte Gebiete* **15**: 195–201.

Minnotte, M. C. (1997). Nonparametric testing of the existence of modes, *Ann. Statist.* **25**: 1646–1660.

Minnotte, M. C. and Scott, D. W. (1993). The mode tree: a tool for visualization of nonparametric density features, *J. Comp. Graph. Statist.* **2**: 51–68.

Mira, A., Møller, J. and Roberts, G. O. (2001). Perfect slice sampler, *J. Roy. Statist. Soc. B* **63**: 593–606.

Møller, J. (1989). On the rate of convergence of spatial birth-and-death processes, *Ann. Inst. Statist. Math.* **41**: 565–581.

Møller, J. (1999). Markov chain Monte Carlo and spatial point processes, *in* O. E. Barndorff-Nielsen, W. S. Kendall and M. N. M. van Lieshout (eds), *Stochastic Geometry: Likelihood and Computations*, Chapman and Hall/CRC, pp. 141–172.

Møller, J. (2001a). A comparison of spatial point process models in epidemiological applications, *in* P. J. Green, N. L. Hjort and S. Richardson (eds), *Highly Structured Stochastic Systems*, Oxford University Press, Oxford. To appear.

Møller, J. (2001b). A review of perfect simulation in stochastic geometry, *IMS Lecture Notes - Monograph Series: Volume 37*, pp. 333–355.

Møller, J. and Nicholls, G. K. (1999). Perfect simulation for sample-based inference, *Technical report*, Department of Mathematics, University of Auckland.

Møller, J. and Skare, O. (2001). Bayesian image analysis with coloured Voronoi tessellations and a view to applications in reservoir modelling, *Technical report*, Department of Mathematical Sciences, Aalborg University.

Møller, J. and Waagepetersen, R. P. (2001). Simulation based inference for spatial point processes, *in* M. B. Hansen and J. Møller (eds), *Spatial Statistics and Computational Methods*, Lecture Notes in Statistics, Springer-Verlag. To appear.

Møller, J., Syversveen, A. and Waagepetersen, R. P. (1998). Log Gaussian Cox processes, *Scandinavian Journal of Statistics* **25**: 451–482.

Muise, R. and Smith, C. (1992). Nonparametric minefield detection and localization, *Technical report*, Naval Surface Warfare Center, Coastal Systems Station.

Müller, D. W. and Sawitzki, G. (1991). Excess mass estimates and tests for multimodality, *J. Amer. Statist. Assoc.* **86**: 738–746.

Neyman, J. and Scott, E. L. (1958). Statistical approach to problems of cosmology, *Journal of the Royal Statistical Society B* **20**: 1–43.

Nicholls, G. K. (1998). Bayesian image analysis with Markov chain Monte Carlo and coloured continuum triangulation models, *J. Roy. Statist. Soc. B* **60**: 643–659.

Nowak, R. D. and Kolaczyk, E. D. (2000). A bayesian multiscale framework for poisson inverse problems, *IEEE Transactions on Information Theory* **46**: 1811–1825.

Obled, C. and Creutin, J. D. (1986). Some developments in the use of empirical orthogonal functions for mapping meteorological fields, *J. Climate and Applied Meteorology* **25**: 1189–1204.

Ogata, Y. and Katsura, K. (1988). Likelihood analysis of spatial inhomogeneity for marked point patterns, *Ann. Inst. Statist. Math.* **40**: 29–39.

Ohser, J. and Mücklich, F. (2000). *Statistical Analysis of Microstructures in Materials Science*, Wiley, New York.

Okabe, A., Boots, B., Sugihara, K. and Chiu, S.-N. (2000). *Spatial tessellations: Concepts and applications of Voronoi diagrams*, 2nd edn, Chichester: Wiley.

Pascal-Marqui, R. M., Michel, C. and Lehmann, D. (1994). Low resolution electromagnetic tomography. a new method for localizing electrical activity., *International Journal of Psychophysiology* **18**: 49–65.

Penttinen, A. and Stoyan, D. (1989). Statistical analysis for a class of line segment processes, *Scan. J. Statist.* **16**: 153–168.

Phillips, D. B. and Smith, A. F. M. (1994). Bayesian faces via hierarchical template modeling, *J. Amer. Statist. Assoc.* **89**: 1151–1163.

Preston, C. J. (1976). *Random fields*, Springer Verlag.

Preston, C. J. (1977). Spatial birth-and-death processes, *Bull. Int. Statist. Inst.* **46**: 371–391.

Propp, J. G. and Wilson, D. B. (1996). Exact sampling with coupled Markov chains and applications to statistical mechanics, *Random Structures and Algorithms* **9**: 223–252.

Protassov, R., van Dyk, D. A., Connors, A., Kashyap, V. and Siemiginowska, A. (2001). Statistics: Handle with care – detecting multiple model components with the likelihood ratio test, *The Astrophysical Journal* p. submitted.

Raftery, A. E., Madigan, D. and Hoeting, J. A. (1997). Bayesian model averaging for linear regression models, *Journal of the American Statistical Association* **92**: 179–191.

Reiss, R. D. (1993). *A course on point processes*, Springer Verlag.

Richardson, S. (2001). Spatial models in epidemiological applications, *in* P. J. Green, N. L. Hjort and S. Richardson (eds), *Highly Structured Stochastic Systems*, Oxford University Press, Oxford. To appear.

Richardson, S. and Green, P. J. (1997). On Bayesian analysis of mixtures with unknown number of components, *Journal of the Royal Statistical Society* **59**: 731–792.

Richardson, W. H. (1972). Bayesian-based iterative methods of image restoration, *Journal of the Optical Society of America* **62**: 55–59.

Ripley, B. D. (1976). The second-order analysis of stationary point processes, *Journal of Applied Probability* **13**: 255–266.

Ripley, B. D. (1977). Modelling spatial patterns (with discussion), *J. Roy. Statist. Soc. B* **39**: 172–212.

Ripley, B. D. (1979). On tests of randomness for spatial point patterns, *J. Roy. Statist. Soc. B* **41**: 368–374.

Ripley, B. D. (1981). *Spatial Statistics*, Wiley.

Ripley, B. D. (1988). *Statistical Inference for Spatial Processes*, Cambridge: Cambridge University Press.

Robbins, C. S., Bystrak, D. A. and Geissler, P. H. (1986). he breeding bird survey:

its first fifteen years, 1965-1979, *Technical report*, USDOI, Fish and Wildlife Service Resource Publication 157. Washington, D.C.

Robert, C. P. and Casella, G. (1999). *Monte Carlo statistical methods*, New York: Springer.

Roberts, G. O. (1996). Markov chain concepts related to sampling algorithms., *Markov Chain Monte Carlo in Practice* (Editors: W. R. Gilks, S. Richardson, and D. J. Spiegelhalter), Chapman & Hall, London.

Roberts, G. O. and Tweedie, R. L. (1996). Exponential convergence of Langevin diffusions and their discrete approximations, *Bernoulli* **2**: 341–363.

Rossky, P. J., Doll, J. D. and Friedman, H. L. (1978). Brownian dynamics as smart Monte Carlo simulation, *Journal of Chemical Physics* **69**: 4628–4633.

Royle, J. A. and Wikle, C. K. (2001). Large-scale modeling of breeding bird survey data, Under review.

Royle, J. A., Link, W. A. and Sauer, J. R. (2001). Statistical mapping of count survey data, *in* J. M. Scott, P. J. Heglund, M. Morrison, M. Raphael, J. Haufler and B. Wall (eds), *Predicting Species Occurrences: Issues of Scale and Accuracy*, Covello, CA: Island Press. (to appear).

Rue, H. and Hurn, M. (1999). Bayesian object identification, *Biometrika* **86**: 649–660.

Ruelle, D. (1969). *Statistical mechanics*, New York: Wiley.

Sambridge, M. (1999a). Geophysical inversion with a neighbourhood algorithm - I. search a parameter space, *Geophys. J. Int.*

Sambridge, M. (1999b). Geophysical inversion with a neighbourhood algorithm - II. appraising the ensemble, *Geophys. J. Int.*

Santaló, L. (1976). *Integral Geometry and Geometric Probability*, Addison–Wesley, Reading, MA.

Sauer, J. R., Pendleton, G. W. and Orsillo, S. (1995). Mapping of bird distributions from point count surveys, *in* C. J. Ralph, J. R. Sauer and S. Droege (eds), *Monitoring Bird Populations by Point Counts*, USDA Forest Service, Pacific Southwest Research Station, pp. 151–160.

Sauer, J. R., Peterjohn, B. G. and Link, W. A. (1994). Observer differences in the North American breeding bird survey, *Auk* **111**: 50–62.

Schlattmann, P. and Böhning, D. (1993). Mixture models and disease mapping, *Statistics in Medicine* **12**: 1943–1950.

Schlattmann, P. and Böhning, D. (1997). On Bayesian analysis of mixtures with an unknown number of components. Contribution to a paper by S. Richardson and P.J. Green, *J. R. Stat. Soc., Ser. B* **59**(4): 782–783.

Scott, D. W. (1992). *Multivariate Density Estimation: Theory, Practice and Visualization*, New York: Wiley.

Serra, J. (1982). *Image analysis and mathematical morphology*, London: Academic Press.

Shepp, L. A. and Vardi, Y. (1982). Maximum likelihood reconstruction for emission tomography, *IEEE Transactions on Image Processing* **2**: 113–122.

Silverman, B. W. (1981). Using kernel density estimates to investigate multimodality, *J. Roy. Statist. Soc. B* **43**: 97–99.

Silverman, J. F. and Cooper, D. B. (1988). Bayesian clustering for unsupervised estimation of surface and texture models, *IEEE Trans. Patt. Anal. Mach. Intell.*

    10: 482–495.

Simonoff, J. (1993). *Smoothing Methods in Statitsics*, Springer Verlag, New York.

Smith, A. F. M. and Roberts, G. O. (1993). Bayesian computation via the Gibbs sampler and related Markov chain Monte Carlo methods, *Journal of the Royal Statistical Society, Series B, Methodological* **55**: 3–23.

Smith, R. L. and Robinson, P. J. (1997). A Bayesian approach to the modeling of spatial-temporal precipitation data, *in* C. Gatsonis, J. S. Hodges, R. E. Kass, R. McCulloch, P. Rossi and N. D. Singpurwalla (eds), *Case Studies in Bayesian Statistics. Vol III*, Springer-Verlag, New York, pp. 237–269.

Snyder, D. L. (1975). *Random Point Processes*, Wiley, New York.

Sourlas, N., van Dyk, D. A., Kashyap, V., Drake, J. and Pease, D. (2002). Bayesian spectral analysis of "MAD" stars., *Statistical Challenges in Modern Astronomy III* (Editors: E. Feigelson and G. Babu), Springer–Verlag, New York.

Spiegelhalter, D. J., Best, N. G. and Carlin, B. P. (1999). Bayesian deviance, the effective number of parameters and the comparison of arbitrarily complex models, *Journal of the Royal Statistical Society*. submitted.

Stanford, D. and Raftery, A. E. (1997). Principal curve clustering with noise, *Technical report*, Department of Statistics, University of Washington, Seattle.

Stein, M. L. (1999). *Interpolation of Spatial Data: Some Theory for Kriging*, New York: Springer-Verlag.

Stephens, D. A. (1994). Bayesian retrospective multiple-changepoint identification, *Appl. Statist.* **43**: 159–178.

Stephens, M. (2000). Bayesian Analysis of Mixture Models with an Unknown Number of Components-an alternative to reversible jump methods, *Annals of Statistics*.

Stoica, R. S., Descombes, X. and Zerubia, J. (2000). A Gibbs point process for road extraction in remotely sensed images, *Technical report*, Research Report 3923, INRIA Sophia Antipolis.

Stoica, R. S., Descombes, X., van Lieshout, M. N. M. and Zerubia, J. (2001). An application of marked point processes to the extraction of linear networks from images, *in* J. Mateu and F. Montes (eds), *Spatial statistics: case studies*, Advances in Ecological Sciences, Volume 13, Southampton: WIT Press.

Stoyan, D. and Stoyan, H. (1994). *Fractals, Random Shapes and Point Fields*, Wiley, Chichester.

Stoyan, D. and Stoyan, H. (2000). Improving ratio estimators of second order point process characteristics, *Scandinavian Journal of Statistics* **27**: 641–656.

Stoyan, D., Kendall, W. S. and Mecke, J. (1995). *Stochastic Geometry and Its Applications*, second edn, Wiley, Chichester.

Strauss, D. J. (1975). A new model for clustering, *Biometrika* **63**: 467–475.

Sun, D., Tsutakawa, R. and Kim, H. (2000). Spatio-temporal interaction with disease mapping, *Statistics in Medicine*. to appear.

Talairach, J. and Tournoux, P. (1988). *Co-planar Stereotaxic Atlas of the Human Brain.*, Thieme Medical Publishers, Stuttgart New York.

Tanner, M. A. (1996). *Tools for Statistical Inference*, New York: Springer-Verlag.

Tanner, M. A. and Wong, W. H. (1987). The calculation of posterior distributions by data augmentation (with discussion), *Journal of the American Statistical Association* **82**: 528–550.

ter Haar Romeny, B. M. (2001). *Front-End Vision and Multiscale Image Analysis*, Dordrecht: Kluwer.

Thomas, M. (1949). A generalization of Poisson's binomial limit for use in ecology, *Biometrika* **36**: 18–25.

Tikhonov, A. N. and Arsenin, V. Y. (1977). *Solutions of Ill-posed problems*, Wiley, Chichester New York.

Titterington, D. M., Smith, A. F. M. and Makov, U. E. (1985). *Statistical Analysis of Finite Mixture Distributions*, New York: John Wiley.

Tzourio, N., Massioui, F. E., Crivello, F., Joliot, M., Renault, B. and Mazoyer, B. (1997). Functional anatomy of human auditory attention studied with PET, *Neuroimage* **5**: 63–77.

van Dyk, D. A. (2002). Hierarchical models, data augmentation, and Markov chain Monte Carlo, *Statistical Challenges in Modern Astronomy III* (Editors: E. Feigelson and G. Babu), Springer–Verlag, New York.

van Dyk, D. A. and Meng, X.-L. (2000). Algorithms based on data augmentation, *Computing Science and Statistics: Proceedings of the 31st Symposium on the Interface* (Editors: M. Pourahmadi and K. Berk), Interface Foundation of North America, Fairfax Station, VA, pp. 230–239.

van Dyk, D. A., Connors, A., Kashyap, V. and Siemiginowska, A. (2001). Analysis of energy spectra with low photon counts via Bayesian posterior simulation, *The Astrophysical Journal* **548**: 224–243.

van Lieshout, M. N. M. (1995). Stochastic geometry models in image analysis and spatial statistics, *Technical report*, CWI tract 108, Amsterdam.

van Lieshout, M. N. M. (2000). *Markov Point Processes and Their Applications*, Imperial College Press, London.

van Lieshout, M. N. M. and Baddeley, A. J. (1995). Markov chain monte carlo methods for clustering of image features, *Proceedings of the 5th IEE International Conference on Image Processing and Its Applications*, IEE Press, pp. 241–245.

van Lieshout, M. N. M. and Baddeley, A. J. (1996). A nonparametric measure of spatial interaction in point patterns, *Statistica Neerlandica* **50**: 344–361.

van Lieshout, M. N. M. and van Zwet, E. W. (2001). Exact sampling from conditional Boolean models with applications to maximum likelihood inference, *Adv. Appl. Prob* **33**: 339–353.

van Lieshout, M. N. M., Molchanov, I. S. and Zuyev, S. A. (2001). Clustering methods based on variational analysis in the space of measures, *Biometrika*. (to appear).

Voronoi, M. G. (1908). Nouvelles applications des paramètres continus à la théorie des formes quadratiques, *J. Reine Angew. Math.* **134**: 198–287.

Wahba, G. (1990). *Spline models for observational data*, Vol. 59, SIAM: Philadelphia.

Waller, L. A., Carlin, B. P., Xia, H. and Gelfand, A. E. (1997a). Hierarchical spatio-temporal mapping of disease rates, *Journal of the American Statistical Association* **92**: 607–617.

Waller, L. A., Carlin, B. P., Xia, H. and Gelfand, A. E. (1997b). Hierarchical spatio-temporal mapping of disease rates, *Journal of the American Statistical Association* **92**: 607–617.

Waller, L. A., Turnbull, B. W., Clark, A. and Nasca, P. (1992). Chronic disease surveillance and testing of clustering of disease and exposure: application to leukemia incidence and TCE-contaminated dump sites in upstate New York, *Environmetrics* **3**: 281–300.

Waller, L. A., Turnbull, B. W., Clark, A. and Nasca, P. (1994). Spatial pattern analyses to detect rare disease clusters, *in* N. Lange, L. Ryan, L. Billard, D. Brillinger, L. Conquest and J. Greenhouse (eds), *Case Studies in Biometry*, New York: John Wiley, pp. 1–23.

Wand, M. P. and Jones, M. C. (1995). *Kernel smoothing*, London: Chapman & Hall.

Whittemore, A., Friend, N., Brown, B. W. and Holly, E. A. (1987). A test to detect clusters of disease, *Biometrika* **74**: 31–35.

Whittle, P. (1986). *Systems in Stochastic Equilibrium*, Wiley, Chichester.

Widom, B. and Rowlinson, J. S. (1970). A new model for the study of liquid-vapor phase transitions, *J. Chem. Phys.* **52**: 1670–1684.

Wikle, C. K. and Cressie, N. A. C. (1999). A dimension reduction approach to space-time Kalman filtering, *Biometrika* **86**: 815–829.

Wikle, C. K., Berliner, L. M. and Cressie, N. A. C. (1998). Hierarchical Bayesian space-time models, *Environmental and Ecological Statistics* **5**: 117–154.

Wikle, C. K., Milliff, R. F., Nychka, D. and Berliner, L. M. (2001). Spatiotemporal hierarchical Bayesian modeling: Tropical ocean surface winds, *J. Amer. Statist. Assoc.* **96**: 382–397.

Wilson, D. B. (2000). How to couple from the past using a read-once source of randomness, *Random Structures and Algorithms* **16**: 85–113.

Wolpert, R. L. and Ickstadt, K. (1998). Poisson/gamma random field models for spatial statistics, *Biometrika* **85**: 251–267.

Wood, A. T. A. and Chan, G. (1994). Simulation of stationary Gaussian processes in $[0, 1]^d$, *Journal of Computational and Graphical Statistics* **3**: 409–432.

Xia, H. and Carlin, B. P. (1998). Spatio-temporal models with errors in covariates: Mapping ohio lung cancer mortality, *Statistics in Medicine* **17**: 2025–2043.

Yaglom, A. M. (1987). *Correlation Theory of Stationary and Related Random Functions I. Basic Results.*, New York: Springer-Verlag.

Yao, Y. C. (1984). Estimation of a noisy discrete-time step function: Bayes and empirical Bayes approaches, *Ann. Statist.* **12**: 1434–1447.

Zellner, A. (1976). Bayesian and non-Bayesian analysis of the regression model with multivariate Student-t error terms, *J. Amer. Statist. Assoc.*

Zia, H., Carlin, B. P. and Waller, L. A. (1997). Hierarchical models for mapping Ohio lung cancer rates, *Environmetrics* **8**: 107–120.

Index

Author Index

Printed and bound by CPI Group (UK) Ltd, Croydon, CR0 4YY

23/10/2024

01778227-0013